SpringerBriefs in Mathematics

Series Editors

Nicola Bellomo, Torino, Italy

Michele Benzi, Pisa, Italy

Palle Jorgensen, Iowa City, USA

Roderick Melnik, Waterloo, Canada

Otmar Scherzer, Linz, Austria

Benjamin Steinberg, New York, NY, USA

Lothar Reichel, Kent, USA

Yuri Tschinkel, New York, NY, USA

George Yin, Detroit, USA

Ping Zhang, Kalamazoo, MI, USA

SpringerBriefs present concise summaries of cutting-edge research and practical applications across a wide spectrum of fields. Featuring compact volumes of 50 to 125 pages, the series covers a range of content from professional to academic. Briefs are characterized by fast, global electronic dissemination, standard publishing contracts, standardized manuscript preparation and formatting guidelines, and expedited production schedules.

Typical topics might include:

- A timely report of state-of-the art techniques
- A bridge between new research results, as published in journal articles, and a contextual literature review
- A snapshot of a hot or emerging topic
- An in-depth case study
- A presentation of core concepts that students must understand in order to make independent contributions

SpringerBriefs in Mathematics showcases expositions in all areas of mathematics and applied mathematics. Manuscripts presenting new results or a single new result in a classical field, new field, or an emerging topic, applications, or bridges between new results and already published works, are encouraged. The series is intended for mathematicians and applied mathematicians. All works are peer-reviewed to meet the highest standards of scientific literature.

Titles from this series are indexed by Scopus, Web of Science, Mathematical Reviews, and zbMATH.

Wojciech M. Kozlowski

Fixed Points of Semigroups of Pointwise Lipschitzian Operators

A Nonexpansive and Asymptotic Approach

 Springer

Wojciech M. Kozlowski
School of Mathematics and Statistics
University of New South Wales
Sydney, NSW, Australia

ISSN 2191-8198 ISSN 2191-8201 (electronic)
SpringerBriefs in Mathematics
ISBN 978-3-032-08868-0 ISBN 978-3-032-08869-7 (eBook)
https://doi.org/10.1007/978-3-032-08869-7

Mathematics Subject Classification: 47H10, 47H20, 47H09, 47J35, 46B20, 47J25, 47J26

© The Editor(s) (if applicable) and The Author(s), under exclusive license to Springer Nature Switzerland AG 2026

This work is subject to copyright. All rights are solely and exclusively licensed by the Publisher, whether the whole or part of the material is concerned, specifically the rights of translation, reprinting, reuse of illustrations, recitation, broadcasting, reproduction on microfilms or in any other physical way, and transmission or information storage and retrieval, electronic adaptation, computer software, or by similar or dissimilar methodology now known or hereafter developed.
The use of general descriptive names, registered names, trademarks, service marks, etc. in this publication does not imply, even in the absence of a specific statement, that such names are exempt from the relevant protective laws and regulations and therefore free for general use.
The publisher, the authors and the editors are safe to assume that the advice and information in this book are believed to be true and accurate at the date of publication. Neither the publisher nor the authors or the editors give a warranty, expressed or implied, with respect to the material contained herein or for any errors or omissions that may have been made. The publisher remains neutral with regard to jurisdictional claims in published maps and institutional affiliations.

Mathematics Subject Classification (2020): Primary: 47H10, 47H20; Secondary: 47H09, 46B20, 47J25, 47J26, 47J35

This Springer imprint is published by the registered company Springer Nature Switzerland AG
The registered company address is: Gewerbestrasse 11, 6330 Cham, Switzerland

If disposing of this product, please recycle the paper.

Preface

This book aims to provide an overview of recent advancements in fixed-point theory for pointwise Lipschitzian semigroups of nonlinear operators, with a particular emphasis on the asymptotic approach. It is important to highlight that, despite this focus, all results presented can also be applied to more classical contexts, including asymptotic and eventually nonexpansive semigroups, and, most notably, nonexpansive semigroups. Readers primarily interested in nonexpansive semigroups will find a useful and concise survey of the most important results related to common fixed points for these semigroups, presented from a broader perspective. The book offers a thorough motivation for the exploration of pointwise Lipschitzian mappings and highlights the benefits of the asymptotic approach to semigroups of these mappings from both theoretical and practical perspectives. The author firmly believes that this broader viewpoint can yield new insights into established results and inspire novel research directions.

Melbourne, Australia Wojciech M. Kozlowski
August 2025

Competing Interests The author has no competing interests to declare that are relevant to the content of this manuscript.

Contents

1 **Introduction** 1
   1.1 Why Semigroups of Operators? 1
   1.2 Why Nonlinear? 4
   1.3 Why Pointwise Lipschitzian? 5
   1.4 Why Asymptotic? 7
   1.5 How Is This Book Organised? 12

2 **Existence of Common Fixed Points for Pointwise Lipschitzian Semigroups** 13
   2.1 Common Fixed-Point Existence for Asymptotic Pointwise Contractive Semigroups 13
   2.2 Common Fixed-Point Existence for Asymptotic Pointwise Nonexpansive Semigroups 15
   2.3 Notes 20

3 **Our Toolset for Constructing Common Fixed Points** 23
   3.1 Preliminaries 23
   3.2 Approximate Fixed-Point Sequences 27
   3.3 The Demiclosedness Principle for Asymptotic Pointwise Nonexpansive Semigroups 29
   3.4 Asymptotic Pointwise Nonexpansive Sequences 33
   3.5 Notes 40

4 **Generalised Krasnosel'skii-Mann Iteration Processes** 41
   4.1 Preliminaries 41
   4.2 Weak Convergence of Generalised Krasnosel'skii-Mann Iteration Processes in Uniformly Convex Banach Spaces with the Opial Property 45
   4.3 Weak Convergence of Generalised Krasnosel'skii-Mann Iteration Processes in Uniformly Convex and Uniformly Smooth Banach Spaces 49
   4.4 Strong Convergence of Generalised Krasnosel'skii-Mann Iteration Processes 52

4.5	Special Cases	55
4.6	Notes	58

5 Generalised Ishikawa Iteration Processes 59
- 5.1 Preliminaries 59
- 5.2 Weak Convergence of Generalised Ishikawa Iteration Processes in Banach Spaces with the Opial Property 63
- 5.3 Weak Convergence of Generalised Ishikawa Iteration Processes in Uniformly Convex and Uniformly Smooth Banach Spaces 65
- 5.4 Strong Convergence of Generalised Ishikawa Iteration Processes 68
- 5.5 Notes 70

6 Implicit Iteration Processes 71
- 6.1 Preliminaries 71
- 6.2 Weak Convergence of Implicit Iteration Processes in Uniformly Convex Banach Spaces with the Opial Property 77
- 6.3 Weak Convergence of Implicit Iteration Processes in Uniformly Convex and Uniformly Smooth Banach Spaces 78
- 6.4 Strong Convergence of Implicit Iteration Processes 79
- 6.5 Notes 80

7 Stability of Common Fixed Point Construction Processes 83
- 7.1 Stability of Asymptotic Pointwise Nonexpansive Sequences 83
- 7.2 Stability of Krasnosel'skii-Mann Processes 89
- 7.3 Stability of Ishikawa Processes 91
- 7.4 Stability of Implicit Iteration Processes 93
- 7.5 Notes 94

8 Semigroups of Monotone Operators 95
- 8.1 Introduction 95
- 8.2 Preliminaries 96
- 8.3 Existence of Common Fixed Points 98
- 8.4 Weak-Convergence 103
- 8.5 Strong-Convergence 111
- 8.6 Notes 112

9 Applications and Related Topics 115
- 9.1 Application to Differential Equations and Dynamical Systems 115
- 9.2 The Monotone Case 124
- 9.3 Semigroups Corresponding to Stochastic Processes 129
- 9.4 Beyond Banach Spaces 131

References 133

Index 139

Chapter 1
Introduction

Abstract The introductory chapter of this book seeks to address fundamental questions concerning its subject matter. It is organised into several sections, each titled respectively: "Why semigroups of operators?", "Why nonlinear", "Why pointwise Lipschitzian?", and "Why asymptotic?". In response to these questions, we elucidate the natural occurrence of nonlinear, pointwise Lipschitzian mappings and explain how the common fixed points of semigroups of such mappings are related to the stationary points of dynamical systems generated by these semigroups. We observe that stationary points of a process inherently correspond to its long-term behaviour. From this standpoint, adopting an asymptotic perspective is quite rational, as our primary interest is centred on the evolution of the system over time.

1.1 Why Semigroups of Operators?

The evolution of modern nonlinear functional analysis is particularly driven by its substantial application potential, both within mathematics and across various scientific and technological disciplines. A significant number of these applications are related to fixed point theory for nonlinear mappings, particularly utilising the concepts of common fixed points for semigroups of such mappings; refer to monographs [41, 53, 54, 93, 114]. The theory and discussions in this book highlight the nonlinear aspects of these mappings, but it is important to remember that "nonlinear" merely indicates "not necessarily linear." As we move forward, please note that we use the terms "nonlinear mappings" and "nonlinear operators" interchangeably.

Throughout this book, X will always represent a real Banach space. While it is also possible to consider settings that go beyond Banach spaces—see Sect. 9.4 "Beyond Banach Spaces" for further details—our aim is to maintain a straightforward and consistent vector space framework throughout the text. This is particularly important given the technical complexities inherent in pointwise Lipschitzian asymptotic theory. We prefer to concentrate on these aspects rather than introduce additional challenges by transitioning to more general space settings. For the same reason, we typically assume that X is uniformly convex, although several results can still be derived under various, and sometimes weaker, assumptions, such as uniform convexity in every direction. While uniform convexity is often a satisfactory assumption, given the wide range of significant spaces displaying this property, readers in need may wish to consult the literature for more general results or attempt to generalise specific findings to their particular context.

Let C be a nonempty, bounded, closed, and convex subset of the Banach space X. By J we will denote a subsemigroup of $[0, +\infty)$ with normal addition. We assume that $0 \in J$ and that there exists $t > 0$ such that $t \in J$. This assumption implies immediately that $+\infty$ is a cluster point of J in the sense of the natural topology inherited from $[0, \infty)$. The notation $t \to \infty$ will mean that t tends to infinity over J. Common examples include $J = [0, \infty)$ and ideals of the form $J = \{n\alpha : n = 0, 1, 2, 3, ...\}$ for a given $\alpha > 0$.

Definition 1.1 A one-parameter family $\mathcal{T} = \{T_t : t \in J\}$ of mappings from C into itself is called a strongly continuous semigroup, or simply continuous semigroup, on C if it satisfies the conditions

1. $T_0(x) = x$ for $x \in C$;
2. $T_{t+s}(x) = T_t(T_s(x))$ for $x \in C$ and $t, s \in J$;
3. For each $x \in C$, the mapping $t \mapsto T_t(x)$ is strongly continuous.

For every $t \in J$, let $F(T_t)$ denote the set of all fixed points of the mapping T_t. In other words, $F(T_t) = \{x \in C : T_t(x) = x\}$. Note that $F(T_t)$ may be an empty set. It is important to note that if $x \in F(T_t)$, then x is a periodic point (with period t) for the semigroup $\mathcal{T}$, meaning that $T_{kt}(x) = x$ for every natural k.

Definition 1.2 We define the set of all common fixed points for the mappings in $\mathcal{T}$ as the intersection

$$F(\mathcal{T}) = \bigcap_{t \in J} F(T_t).$$

Common fixed points are often interpreted as the stationary points of the system defined by the semigroup $\mathcal{T}$.

Example 1.1 Note that a semigroup $\mathcal{T} = \{T_t : t \in J\}$ is inherently continuous if J is a discrete set. For instance, this is the case when $J = \{0, 1, 2, 3...\}$. Discrete semigroups underpin numerous computational techniques used in information technology, artificial intelligence, and robotics.

Remark 1.1 It is important to note that if $J = \{0, 1, 2, 3...\}$, then the semigroup $\mathcal{T}$ coincides with the semigroup of iterations $\{T^k : k = 0, 1, 2, 3, ...\}$ of an operator $T = T_1$. It is clear that the set of all common fixed points for the semigroup of iterations $\{T^k : k = 0, 1, 2, 3, ...\}$ of T is identical to the set of all fixed points of T itself. We will utilise this fact to show that the fixed point existence results and the fixed point construction processes for a single nonexpansive operator are specific instances of our broader findings regarding semigroups of operators.

1.1 Why Semigroups of Operators?

Remark 1.2 In Sect. 1.4, we will demonstrate that in certain cases, addressing the common fixed point problem for the semigroup $\{T^k : k = 0, 1, 2, 3, ...\}$ of T may be more intuitive than directly seeking a fixed point of T.

Definition 1.3 A continuous semigroup $\mathcal{T}$ is said to be equicontinuous at zero, or simply equicontinuous, if the family of mappings $\{t \mapsto T_t(x) : x \in C\}$ is equicontinuous at $t = 0$. Specifically, this means, that to every $\varepsilon > 0$ there exists a $\delta > 0$ such that $\sup_{x \in C} \|T_t(x) - x\| < \varepsilon$ for all $t \in [0, \delta] \cap J$.

Scenarios based on continuous semigroups[1] of operators are common in mathematics and its applications. For example, in dynamical systems theory, space X would represent the state space, and the mapping $(t, x) \to T_t(x)$ would model the evolution function of a dynamical system. Here, the parameter t can be interpreted as representing either continuous or discrete time, depending on the nature of the parameter set J. Common fixed points of the semigroup can be seen as stationary points of the system, remaining unchanged during the transformation T_t, for every time $t \in J$.

Example 1.2 We conjecture that a dynamical process can be defined by $T_t(x) = u(x, t)$ where $u(x, \cdot) : [0, \infty) \to C$ is a solution of the Cauchy problem:

$$\begin{cases} u(x, 0) = x \\ \dfrac{\partial u}{\partial t}(x, t) + (I - H_t)(u(x, t)) = 0, \end{cases}$$

where $H_t : C \to C$ are nonexpansive mappings. To verify the conjecture, we need to prove that $\{T_t\}$ is a continuous semigroup. In [68], the author demonstrates that this is the case, and, in addition, that this semigroup is monotone nonexpansive provided each H_t is monotone nonexpansive (X is assumed to be an ordered Banach space). We elaborate on this example in Sect. 9.1.

Many significant dynamical systems are solutions of ordinary differential equations (ODEs) or partial differential equations (PDEs), which may be deterministic or stochastic, even though not all such systems are governed by differential equations (see Example 1.4 below).

Example 1.3 The general form of initial value problems associated with dynamical systems—whose solutions may constitute a continuous semigroup, known as the resolving semigroup—can be expressed as follows:

$$\begin{cases} u(x, 0) = x \\ \dfrac{\partial u}{\partial t}(x, t) = g(u(x, t)). \end{cases}$$

---

[1] In this book, we consistently work within the framework of continuous semigroups. However, the reader should be aware that there are important scenarios where semigroups fail to be continuous. This has led to weaker notions of continuity, such as bi-continuous semigroups see, e.g., [79].

Functions g can be quite complicated; in some applications, they can be discontinuous and may involve partial derivatives of unknown functions u.

The method of investigating such problems by employing associated resolving semigroups is well-established for both linear and nonlinear cases. The use of semigroups of nonlinear operators as resolving semigroups has been a focus of intensive research over recent decades. Considering that the state space X can be infinite-dimensional, these results are applicable to both deterministic and stochastic dynamical systems. In this context, developing algorithms to find common fixed points for such semigroups is closely linked to the challenge of solving stochastic evolution equations, as detailed in [12, 25, 28, 39].

As suggested above, by no means are all continuous semigroups generated by solutions to initial value problems. Let us illustrate this by the following example from the 1971 seminal paper by Crandall and Liggett [27].

Example 1.4 Let $X = C([0, 1], \mathbb{R})$, $C = \{f \in X : 0 \le f(x) \le x, \forall x \in [0, 1]\}$. Define mappings $T_t(f)(x) = \min\{t + f(x), x\}$ for $f \in C$, $t \ge 0$. It is straightforward to demonstrate that $\{T_t\}$ is a continuous semigroup but $T_t(f)$ is differentiable at $t = 0$ only if $f(x) = x$ for all $x \in [0, 1]$.

For further discussion and examples demonstrating how the theory developed in this book applies to these scenarios, please refer to Chap. 9 "Applications and related topics".

1.2 Why Nonlinear?

The theory of linear semigroups, where all operators T_t are linear, is well-established and has been successfully applied across numerous scientific and technological fields. For more detailed information, readers are encouraged to consult standard textbooks on linear semigroup theory such as [35, 36, 89].

This leads us to ask: why focus on the nonlinear case? This question hardly demands an answer, as we observe in both science and daily life that most processes, from weather patterns to stock market fluctuations, do not behave linearly. Nonlinear semigroup theory is particularly important in the study of evolution problems related to dynamical processes. We observe the growing interest in these topics, with nonlinear Markov processes, briefly discussed in Chap. 9, serving as a prominent, though certainly not unique, example. Fixed point theory, sitting in the core of nonlinear functional analysis, fits these nonlinear scenarios perfectly. In the context of semigroups of nonlinear mappings, this theory manifests itself as the study of common fixed points for these semigroups with direct implications for stationary points of nonlinear processes.

1.3 Why Pointwise Lipschitzian?

The book's title includes the term "pointwise Lipschitzian" while the subtitle indicates an asymptotic approach. As such, readers may anticipate an explanation for our emphasis on asymptotic pointwise semigroups.

Following [59], let us introduce formal definitions of pointwise Lipschitzian mappings and pointwise Lipschitzian semigroups of mappings, along with the relevant notational conventions.

Definition 1.4 We say that $T : C \to C$ is a pointwise Lipschitzian mapping if for every $x \in C$ there exists $\alpha(x) > 0$ such that

$$\|T(x) - T(y)\| \leq \alpha(x)\|x - y\| \text{ for all } y \in C. \tag{1.1}$$

We say that T s a pointwise contraction if $\alpha(x) < 1$ for every $x \in C$. If $\alpha(x) \leq 1$ for every $x \in C$, then T is referred to as a pointwise nonexpansive mapping.

It is important to note that the existence of pointwise Lipschitzian mappings is not unusual. For example, drawing on findings from [59, Proposition 2.5] and an earlier result by Kirk [49, Proposition 2.1], we can establish the following facts.

Proposition 1.1 ([49, Proposition 2.1]) *Let A be a bounded open convex subset of a Banach space X and let $T : A \to X$ be continuously Fréchet differentiable on A. Then, T is a pointwise contraction mapping on A if and only if $\|T'_{x_0}\| < 1$ for every $x_0 \in A$.*

Recall that, given a convex open subset A of X and $T : A \to X$, we say that T is continuously Fréchet differentiable on A if the mapping $x \mapsto T'_x$ from A to the space of continuous linear operators on X is continuous. Here, T'_x denotes the Fréchet derivative of T at x, namely

$$T'_x(y) = \lim_{t \to 0} \frac{T(x + ty) - T(x)}{t},$$

provided the limit is uniform for all y with $\|y\| = 1$.

By applying Proposition 1.1 to the operator $\frac{1}{\gamma}T$, we can easily demonstrate the following result.

Proposition 1.2 *Let A be a bounded open convex subset of a Banach space X and let $T : A :\to X$ be continuously Fréchet differentiable on A. Given $x \in A$, assume that there exists a $\gamma > 0$ such that $\|T'_x\| < \gamma$. Then, there exists a positive number $\alpha(x) < 1$ such that for every $y \in C$,*

$$\|T(x) - T(y)\| \leq \gamma\alpha(x)\|x - y\|.$$

From Proposition 1.2 we immediately derive the following result.

Corollary 1.1 Let A be a bounded open convex subset of a Banach space X and let $T : A :\to X$ be continuously Fréchet differentiable on A. Then, T is a pointwise Lipschitzian mapping.

The preceding corollary indicates that the investigation of pointwise Lipschitzian semigroups, defined below, encompasses a broad range of applications. Additionally, it is important to note that the characterisation of the pointwise Lipschitzian condition through Fréchet differentiability serves merely as an example and does not establish necessary conditions.

Definition 1.5 A one-parameter family $\mathcal{T} = \{T_t : t \in J\}$ of mappings from C into itself is called a pointwise Lipschitzian semigroup on C if it satisfies the following conditions:

1. $\mathcal{T}$ is a continuous semigroup on C;
2. For each $t \in J$, T_t is a pointwise Lipschitzian mapping, meaning there exists a function $\alpha_t : C \to [0, \infty)$ such that

$$\|T_t(x) - T_t(y)\| \leq \alpha_t(x)\|x - y\| \text{ for all } x, y \in C.$$

Example 1.5 Let us consider the following variant of Example 1.2:

$$\begin{cases} u(x, 0) = x \\ \dfrac{\partial u}{\partial t}(x, t) + (I - H)(u(x, t)) = 0, \end{cases} \tag{1.2}$$

where $H : C \to C$ is continuously Fréchet differentiable on an open set A containing C. By direct calculation one can verify that the operator $G(x) := H(x) - x$ is an infinitesimal generator of the nonlinear semigroup of operators $T_t(x) = u(x, t)$. Recall that this means that

$$G(x) = \lim_{t \to 0^+} \frac{1}{t}\left(x - T_t(x)\right).$$

Although no general results exist for all nonlinear operators, there are specific cases where the continuous Fréchet differentiability of G implies that each T_t is also continuously Fréchet differentiable. Consequently, in such cases, by Corollary 1.1, the class $\{T_t\}$ forms a pointwise Lipschitzian semigroup. For example, refer to [1] for the relevant discussion on the holomorphic case; see also [93]. In Sect. 9.1 we apply different methods to show when $\{T_t\}$, corresponding to the initial value problem (1.2), forms a pointwise Lipschitzian semigroup and an asymptotic pointwise nonexpansive semigroup (Theorem 9.3).

1.4 Why Asymptotic?

The subtitle of the book suggests an asymptotic perspective, leading readers to expect an explanation of our focus on asymptotic semigroups and how they relate to nonexpansive and pointwise Lipschitzian semigroups. To make our discussion more precise, let us start with the formal definitions of various types of semigroups under consideration in this book.

Definition 1.6 A pointwise Lipschitzian semigroup $\mathcal{T}$ is called asymptotic pointwise nonexpansive if $\limsup_{t\to\infty} \alpha_t(x) \leq 1$ for every $x \in C$. The class of all asymptotic pointwise nonexpansive semigroups on C will be denoted by $\mathcal{APNS}(C)$.

Definition 1.7 A pointwise Lipschitzian semigroup $\mathcal{T}$ is called an asymptotic nonexpansive semigroup if it is not "pointwise". This specifically means that each function $\alpha_t(\cdot)$ used in (1.1) is constant over the set C. Denoting these constants by $\{\alpha_t\}_{t \in J}$, our asymptotic nonexpansive condition becomes $\limsup_{t\to\infty} \alpha_t \leq 1$. The class of all asymptotic nonexpansive semigroups on C will be denoted by $\mathcal{ANS}(C)$.

Definition 1.8 A pointwise Lipschitzian semigroup $\mathcal{T}$ is called pointwise eventually nonexpansive if, for every $x \in C$, there exists $t_x \in J$ such that for all $t \in J$ with $t \geq t_x$, the condition $\alpha_t(x) = 1$ holds. The class of all pointwise nonexpansive semigroups on C will be denoted by $\mathcal{PNE}(C)$.

Definition 1.9 A pointwise Lipschitzian semigroup $\mathcal{T}$ is called eventually nonexpansive if there exists $t_0 \in J$ such that for all $t \in J$ with $t \geq t_0$, the condition $\alpha_t(x) = 1$ holds for all $x \in C$.

Definition 1.10 A pointwise Lipschitzian semigroup $\mathcal{T}$ is called nonexpansive if each operator $T_t \in \mathcal{T}$ is a nonexpansive mapping. This means that $\alpha_t(x) = 1$ for every $x \in C$ and all $t \in J$. The class of all nonexpansive semigroups on C will be denoted by $\mathcal{NE}(C)$.

Note that the class of pointwise Lipschitzian semigroups encompasses all other types of semigroups and that the nonexpansive semigroups belong to all other classes. Similarly, we define the contraction variants of pointwise Lipschitzian semigroups.

Definition 1.11 A pointwise Lipschitzian semigroup $\mathcal{T}$ is called asymptotic pointwise contractive if $\limsup_{t\to\infty} \alpha_t(x) < 1$ for every $x \in C$. The class of all asymptotic pointwise contractive semigroups on C will be denoted by $\mathcal{APCS}(C)$.

Definition 1.12 A pointwise Lipschitzian semigroup $\mathcal{T}$ is called an asymptotic contractive semigroup if it is not "pointwise". This specifically means that each function $\alpha_t(\cdot)$ used in (1.1) is constant over the set C. Denoting these constants by $\{\alpha_t\}_{t \in J}$, our asymptotic contractive condition becomes $\limsup_{t\to\infty} \alpha_t < 1$. The class of all asymptotic contractive semigroups on C will be denoted by $\mathcal{ACS}(C)$.

At this point, it is important to note that the term "asymptotic" has various interpretations in mathematics, including within fixed point theory. Although a comparative analysis of these contexts would be interesting, this book will consistently adhere to the definitions provided above. For semigroups of iterations $\{T^k : k = 0, 1, 2, 3, ...\}$ of a Lipschitzian operator T, this approach was pioneered by Goebel and Kirk [40]. Subsequently, Kirk and Xu expanded this methodology to encompass the pointwise Lipschitzian case in [50]. Kozlowski further extended it in [59] to cover pointwise Lipschitzian semigroups. Since then, a considerable body of research has evolved around these concepts, highlighting the significance of the question posed in the title of this section: "Why pointwise asymptotic?"

In his influential 1965 work [17], Browder demonstrated that any commuting family of nonexpansive mappings acting within a given bounded closed convex subset C of a uniformly convex Banach space X has a common fixed point. Recall that the modulus of convexity $\delta(\varepsilon)$ of a Banach spaces X is defined for each $\varepsilon \in [0, 2]$ by

$$\delta(\varepsilon) = \inf \left\{ 1 - \left\| \frac{x+y}{2} \right\| : \|x\| \leq 1, \|y\| \leq 1, \|x - y\| \geq \varepsilon \right\}.$$

The Banach space X is termed uniformly convex if $\delta(\varepsilon) > 0$ for every $\varepsilon > 0$. By applying Browder's result to the semigroup $\mathcal{T} = \{T_t\}_{t \in J}$ of nonexpansive operators, we can conclude that, under the specified conditions on X and C, the semigroup $\mathcal{T}$ possesses a common fixed point. While this result is both elegant and critically important, it often poses challenges in practical applications due to the difficulty in selecting an appropriate set C where all $T_t \in \mathcal{T}$ are nonexpansive. The primary advantage of the asymptotic approach lies in its ability to significantly relax the requirement for all operators in $\mathcal{T}$ to be nonexpansive. To illustrate this point, let us the introduce the following simple example.

Example 1.6 Imagine that a researcher describes an experimental process using the semigroup $\mathcal{T}$ of operators defined by $T_t(x) = x^{2^{-t}}$, where $t \in [0, +\infty)$, and $x \geq x_0 = 0.001$. It is easy to verify that $\mathcal{T}$ is a continuous semigroup. Our researcher believes—correctly—that there must exist a positive number less than 100 that serves as a stationary point for this process, that is, a common fixed point of $\mathcal{T}$. However, it is unclear how Browder's theorem can be applied directly, as not all of the operators in $\mathcal{T}$ are everywhere nonexpansive. By applying the Mean Value Theorem, we find that for any $x \geq x_0$ and any $y \geq x_0$,

$$|T_t(x) - T_t(y)| \leq \alpha_t(x)|x - y|,$$

where $\alpha_t(x) = \max\{2^{-t}x^{2^{-t}-1}, 2^{-t}x_0^{2^{-t}-1}\}$, which tends to zero as $t \to \infty$. This means that $\mathcal{T}$ is an asymptotic pointwise contractive semigroup on $[0.001, 100]$. The question arises of whether there exists a general theorem on the existence of common fixed points for such semigroups that our researcher could utilise to guarantee the existence of a stationary point for the process described by $\mathcal{T}$. Indeed, we will prove in Chap. 2 that this is the case. Furthermore, in the following chapters, we will explore several algorithms to construct such points.

1.4 Why Asymptotic?

Interestingly, general theorems on the existence of common fixed points for the asymptotic pointwise nonexpansive semigroups can be used to prove the existence of a fixed point of a single operator. To illustrate this point, let us revisit an example provided in the original paper by Goebel and Kirk [40].

Example 1.7 Let C denote the unit ball in the Hilbert space l^2 and let T be defined as follows,
$$T : (x_1, x_2, x_3, \cdots) \mapsto (0, x_1^2, A_2 x_2, A_3 x_3, \cdots),$$
where $\{A_i\}$ is a sequence of numbers such that $0 < A_i < 1$ and $\prod_{i=2}^{\infty} A_i = \frac{1}{2}$. It is easy to see that T is not nonexpansive in C, however the semigroup $\{T^k\}$ of iterations of T is asymptotic pointwise nonexpansive in C because for every $x, y \in C$,
$$\|T^k(x) - T^k(y)\| \leq \alpha_k \|x - y\|,$$
where $\alpha_k = 2 \prod_{i=2}^{k} A_i$ for $k = 2, 3, \cdots$, and $\lim_{k \to \infty} \alpha_k = 1$. Thus, using the original Goebel-Kirk Theorem [40, Theorem 1] or our asymptotic results from Chap. 2, we are be able to demonstrate the existence of a common fixed point for $\{T^k\}$, which will also serve as fixed point of T itself. It is actually quite clear that $(0, 0, 0, \cdots)$ is such a fixed point. The issue, which the authors of this paper did not expand upon, is to understand why this fixed point was not identified using the classical Browder's Theorem. The choice of the unit ball as the set C is obviously a limiting factor. If, instead, we consider the mapping $T : D \to D$, where the set $D = \{x \in C : |x_1| \leq \frac{1}{2}\}$, it becomes apparent that T is nonexpansive in D. Since D is bounded, closed and convex, it follows from Browder's Theorem, that T must have a fixed point in D. The critical point is that, a priori, it was unclear which set should be selected. This straightforward example illustrates the power of the asymptotic approach, which can provide solutions in scenarios where classical methods may fall short.

A curious reader might wonder if there is a broader pattern underlying such examples, or in other words, a formal framework that can help in applying the concepts from the theory of asymptotic pointwise semigroups. In response, we provide a useful characterisation of asymptotic pointwise nonexpansive semigroups of Fréchet differentiable operators in Banach spaces.

Proposition 1.3 *Let $\mathcal{T}$ be a pointwise Lipschitzian semigroup on C, where C is a nonempty, bounded, closed, and convex subset of a Banach space X. Assume that all mappings $T_t \in \mathcal{T}$ are continuously Fréchet differentiable on an open convex set A containing C then $\mathcal{T}$ is asymptotic pointwise contractive on C if and only if for each $x \in C$, $\limsup_{t \to \infty} \|(T_t)'_x\| < 1$.*

Proof Given $x \in C$, assume that $\limsup_{t \to \infty} \|(T_t)'_x\| < 1$. There exist then a positive $\gamma < 1$ and a $t_0 > 0$ such that $\|(T_t)'_x\| < \gamma$ for every $t > t_0$. Then, by Proposition 1.2, for every $t > t_0$ there exists a positive number $\alpha_t(x) < 1$ such that for every $y \in C$,
$$\|T_t(x) - T_t(y)\| \leq \gamma \alpha_t(x) \|x - y\|.$$

For any $0 \leq t \leq t_0$ there exists $0 < \gamma_t$ such that $\|(T_t)'_x\| < \gamma_t$. It follows from Proposition 1.2 again that there exists a $0 < \beta_t(x) < 1$ such that for every $y \in C$,

$$\|T_t(x) - T_t(y)\| \leq \gamma_t \beta_t(x) \|x - y\|.$$

Define

$$a_t(x) = \begin{cases} \gamma_t \beta_t(x), & \text{for } t \leq t_0, \\ \gamma \alpha_t(x), & \text{for } t > t_0, \end{cases}$$

and observe that for every $y \in C$,

$$\|T_t(x) - T_t(y)\| \leq a_t(x) \|x - y\|, \qquad (1.3)$$

and that

$$\limsup_{t \to \infty} a_t(x) = \gamma \limsup_{t \to \infty} \alpha_t(x) \leq \gamma < 1.$$

Therefore, $\mathcal{T}$ is an asymptotic pointwise contractive semigroup on C. To prove the other direction, assume that $\mathcal{T}$ is such a semigroup satisfying (1.3), where $\limsup_{t \to \infty} a_t(x) < 1$. There exist then a positive $\gamma < 1$ and a $t_0 > 0$ such that $a_t(x) < \gamma$ for every $t > t_0$, which implies that

$$\|T_t(x) - T_t(y)\| < \gamma \|x - y\|$$

for every $y \in C$. By applying Proposition 1.1 to the operator $\frac{1}{\gamma} T$, we deduce that the Fréchet derivative satisfies the inequality $\|(T_t)'_x\| < \gamma$ for every $t > t_0$. This implies that $\limsup_{t \to \infty} \|(T_t)'_x\| < 1$, as claimed. $\square$

Likewise, we can characterise asymptotic pointwise nonexpansive semigroups based on the asymptotic behaviour of the Fréchet derivatives of their members, as demonstrated in the following result. As before, the proof utilises Propositions 1.1 and 1.2. However, due to certain subtleties, it is beneficial to present the proof in full detail.

Proposition 1.4 *Let $\mathcal{T}$ be a pointwise Lipschitzian semigroup on C, where C is a nonempty, bounded, closed and convex subset of a Banach space X. Assume that all mappings $T_t \in \mathcal{T}$ are continuously Fréchet differentiable on an open convex set A containing C then $\mathcal{T}$ is asymptotic pointwise nonexpansive on C if and only if for each $x \in C$, $\limsup_{t \to \infty} \|(T_t)'_x\| \leq 1$.*

Proof Let us fix a strictly decreasing sequence $0 < \varepsilon_n \downarrow 0$. Assume first that $\limsup_{t \to \infty} \|(T_t)'_x\| \leq 1$. Then, for every $n \in \mathbb{N}$ there exists a $t_n > 0$ such that for every $t > t_n$,

$$\|(T_t)'_x\| < 1 + \varepsilon_n. \qquad (1.4)$$

1.4 Why Asymptotic?

We can assume that the sequence $t_n \uparrow +\infty$. Define $\delta_t(x) = 1 + \varepsilon_n$ for $t_n < t \le t_{n+1}$, where by convention we set $t_0 = 0$. Then, by applying (1.4), we have

$$\|(T_t)'_x\| < \delta_t(x).$$

It follows from Proposition 1.2 that there exists a positive number $\alpha_t(x) < 1$ such that for every $y \in C$,

$$\|T_t(x) - T_t(y)\| \le \delta_t(x)\alpha_t(x)\|x - y\|.$$

Let $a_t(x) = \delta_t(x)\alpha_t(x)$. Since $\delta_t(x) \to 1$ as $t \to \infty$, combined with the fact that $\alpha_t(x) < 1$ for every $t \ge 0$, we conclude that $\limsup_{t \to \infty} a_t(x) \le 1$, which implies that $\mathcal{T}$ is asymptotic pointwise nonexpansive on C.

To prove the other direction of our hypothesis, let us assume that $\mathcal{T}$ is such a semigroup satisfying

$$\|T_t(x) - T_t(y)\| \le a_t(x)\|x - y\|, \tag{1.5}$$

where $\limsup_{t \to \infty} a_t(x) \le 1$. Then, for every $n \in \mathbb{N}$ there exists $t_n > 0$ such that for every $t > t_n$, we have $a_t(x) < 1 + \varepsilon_{n+1}$. We can assume that $t_n \uparrow +\infty$. Let $t_0 = 0$. By (1.5),

$$\left\|\frac{1}{1+\varepsilon_n}T_t(x) - \frac{1}{1+\varepsilon_n}T_t(y)\right\| < \gamma_n \|x - y\|,$$

holds for $t_n < t \le t_{n+1}$, where $\gamma_n = \frac{1+\varepsilon_{n+1}}{1+\varepsilon_n}$. Since $\gamma_n < 1$, it follows from Proposition 1.1 that

$$\|(T_t)'_x\| < 1 + \varepsilon_n,$$

for $t_n < t \le t_{n+1}$. Hence, $\limsup_{t \to \infty} \|(T_t)'_x\| \le \lim_{n \to \infty}(1 + \varepsilon_n) = 1$, completing the proof. $\square$

We emphasise that the characterisations offered in Propositions 1.3 and 1.4 are intended to illustrate the existence of general criteria; however, they do not encompass all possible scenarios.

We conclude this section by noting that the stationary points of a process relate, by their nature, to its long-term behaviour. From this perspective, the asymptotic approach is quite sensible, as our primary interest lies in the system's evolution over time.

1.5 How Is This Book Organised?

The research outlined in this book addresses four primary topics: (1) existence and characteristics of common fixed points, (2) methods for approximating these points constructively, (3) stability of such algorithms, and (4) application to differential equations and dynamical systems.

Chapter 2 surveys key findings on the existence of common fixed points for asymptotic pointwise contractive and asymptotic pointwise nonexpansive semigroups. Additionally, it explores fundamental properties of common fixed point sets.

Chapter 3 introduces several key concepts and technical results essential for proving the convergence of the common fixed point construction processes.

The techniques and tools established in the previous chapters are applied in Chap. 4 to adapt the Krasnosel'skii-Mann method to the setting of pointwise Lipschitzian semigroups and discuss its convergence to a common fixed point.

These results are then extended in Chap. 5 to a two-step generalisation of the Krasnosel'skii-Mann process, known as the Ishikawa process.

In Chap. 6 we consider the convergence of implicit iteration processes in the context of pointwise Lipschitzian semigroups.

In Chap. 7, we examine the stability of these algorithms concerning computational errors.

Chapter 8 is dedicated to an important class of monotone pointwise Lipschitzian semigroups and their applications.

The final Chap. 9 concentrates on applying this theory to deterministic and stochastic differential equations and related dynamical systems, emphasising their importance in scientific and technological applications. The last section of this chapter offers a brief reference to the parallel common fixed point results in metric, hyperbolic, and modular function spaces.

To ensure a smooth reading experience, we aim to minimise references to results within the main text, unless necessary. Most chapters include a "Notes" section that provides references to the literature and additional commentaries.

Chapter 2
Existence of Common Fixed Points for Pointwise Lipschitzian Semigroups

Abstract Let X be a uniformly convex Banach space and $C \subset X$ be bounded, closed, and convex. In this chapter, we prove that, in this context, every asymptotic pointwise contractive semigroup possesses a unique common fixed point in C, which can be constructed by taking the strong limit of orbits. Furthermore, we demonstrate that every asymptotic pointwise nonexpansive semigroup has at least one fixed point in C and that the set of all common fixed points is closed and convex.

2.1 Common Fixed-Point Existence for Asymptotic Pointwise Contractive Semigroups

A concept of types plays an important role in proofs of many fixed point results. A typical definition of a type is based on a given sequence. For our purpose we need to adapt this definition to one-parameter semigroups of mappings.

Definition 2.1 Let C be a nonempty, convex, and bounded subset of X. A function $\varphi : C \to [0, \infty]$ is called a type if there exists a one-parameter family $\{z_t\}_{t \in J}$ of elements of C such that for any $y \in C$,

$$\varphi(y) = \limsup_{t \to \infty} \|z_t - y\| = \inf_{M > 0} \left(\sup_{t \in J_M} \|z_t - y\| \right),$$

where $J_M = J \cap [M, \infty)$.

Remark 2.1 Observe that a type φ is a convex and bounded function on a convex and bounded set C. It is easy to show that, for any $\alpha > 0$, the set $U_\alpha = \varphi^{-1}([0, \alpha])$ is a convex and closed subset of C. It follows from Mazur's Theorem that the set U_α is closed in the weak topology, which implies that the type φ is weak lower semicontinuous on C.

Remark 2.2 Let X be a uniformly convex Banach space, which implies, by the Milman-Pettis Theorem, that X is reflexive. Let $C \subset X$ be nonempty, convex, and bounded. Then, by the Eberlein-Šmulian Theorem, the set C is weakly compact. In

Remark 2.1, we observed that any type φ on C is lower semicontinuous with respect to the weak topology. Consequently, φ must attain its minimum in C; that is, there exists an element $z \in C$ such that $\varphi(z) = \min\{\varphi(y) : y \in C\}$. This property of types defined in uniformly convex Banach spaces will play a prominent role in proving several of our results.

Theorem 2.1 *Assume that X is a uniformly convex Banach space. Let $C \subset X$ be a nonempty, bounded, closed, and convex set. Suppose $\mathcal{T}$ is an asymptotic pointwise contractive semigroup on C. Then, $\mathcal{T}$ possesses a unique common fixed point $z \in C$. Furthermore, for every $x \in C$,*

$$\lim_{t \to \infty} \|T_t(x) - z\| = 0. \tag{2.1}$$

Proof Fix $x \in C$ and define the type function φ by

$$\varphi(y) = \limsup_{t \to \infty} \|T_t(x) - y\|.$$

As explained in Remark 2.2, since φ is weakly lower semicontinuous on a weakly compact set C then it attains its minimum at a point $z \in C$. Observe that

$$\varphi(T_s(y)) = \limsup_{t \to \infty} \|T_t(x) - T_s(y)\| = \limsup_{t \to \infty} \|T_{s+t}(x) - T_s(y)\| \tag{2.2}$$

$$\leq \limsup_{t \to \infty} \alpha_s(y)\|T_t(x) - y\| = \alpha_s(y)\varphi(y)$$

holds for any given $y \in C$ and $s \in J$. By applying (2.2) to z and leveraging the minimality of z, we obtain

$$\varphi(z) \leq \varphi(T_s(z)) \leq \alpha_s(z)\varphi(z). \tag{2.3}$$

By passing with s to infinity in (2.3) we get

$$\varphi(z) \leq \alpha(z)\varphi(z),$$

where $\alpha(z) = \limsup_{t \to \infty} \alpha_t(x) < 1$, which implies that $\varphi(z) = 0$. Then, by (2.2), for every $s \in J$,

$$\lim_{t \to \infty} \|T_t(x) - T_s(z)\| = 0. \tag{2.4}$$

By applying (2.4) with $s = 0$ we conclude that

$$\lim_{t \to \infty} \|T_t(x) - z\| = 0. \tag{2.5}$$

From (2.4) and (2.5) it follows immediately that $T_s(z) = z$. Since $s \in J$ was chosen arbitrarily, it follows that $z \in F(\mathcal{T})$, as claimed. To prove the uniqueness, let us take another $w \in F(\mathcal{T})$. Therefore, for each $t \in J$,

$$\|z - w\| = \|T_t(z) - T_t(w)\| \leq \alpha_t(z)\|z - w\|. \tag{2.6}$$

By passing with t to infinity in (2.6), we obtain

$$\|z - w\| \leq \alpha(z)\|z - w\|.$$

Since $\alpha(z) < 1$, this inequality implies that $w = z$. The proof of the theorem is complete. □

Remark 2.3 Theorem 2.1 is noteworthy not only for establishing the existence of a common fixed point z for an asymptotic pointwise contractive semigroup but also for providing a method, defined in (2.1), to construct z as the strong limit of orbits $\{T_t(x)\}$, for any initial point $x \in C$. When the semigroup $\mathcal{T}$ is a semigroup of iterates $\{T^k\}$ of an operator T, this limit becomes $\lim_{k \to \infty} T^k(x)$. Readers may recognise this as the Picard iteration process, utilised in the Banach Fixed Point Principle for finding a fixed point of a contraction T, and known as the method of successive approximation for many centuries prior.

By combining Theorem 2.1 with Proposition 1.3, we immediately obtain the following fixed point result, expressed in a form convenient for the applications.

Theorem 2.2 *Assume that X is a uniformly convex Banach space. Let $C \subset X$ be a nonempty, bounded, closed, and convex set. Suppose $\mathcal{T}$ is a pointwise Lipschitzian semigroup on C. Furthermore, assume that every mapping $T_t \in \mathcal{F}$ is continuously Fréchet differentiable on an open convex set A containing C. Assume that*

$$\limsup_{t \to \infty} \|(T_t)'_x\| < 1, \tag{2.7}$$

holds for every $x \in C$. Then, $\mathcal{T}$ has a unique common fixed point $z \in C$. Furthermore, for every $x \in C$,

$$\lim_{t \to \infty} \|T_t(x) - z\| = 0. \tag{2.8}$$

Returning to Example 1.6, we observe that our researcher can directly apply Theorem 2.2 to conclude that the process defined by the mapping T must have a unique stationary point in the interval $[0.001, 100]$. The formula (2.8) indicates a method for constructing this unique fixed point.

2.2 Common Fixed-Point Existence for Asymptotic Pointwise Nonexpansive Semigroups

The following result generalises the findings of Bruck [19] regarding nonexpansive mappings.

Lemma 2.1 *Let X be a uniformly convex Banach space, and let $C \subset X$ be nonempty, bounded, closed, and convex. There exists a strictly increasing, convex, continuous function $\gamma_2 : [0, \infty) \to [0, \infty)$ with $\gamma_2(0) = 0$, depending only on the diameter of (C), such that for every pointwise Lipschitzian mapping $T : C \to C$, every $c \in [0, 1]$ the inequality*

$$\gamma_2\left(\frac{\|T(cx + (1-c)y) - cT(x) - (1-c)T(y)\|}{\alpha(cx + (1-c)y)}\right)$$
$$\leq \|x - y\| - \frac{\|T(x) - T(y)\|}{\alpha(cx + (1-c)y)} \qquad (2.9)$$

holds for every $x, y \in C$. Recall that, given an $f \in C$, $\alpha(f)$ is such that

$$\|T(f) - T(g)\| \leq \alpha(f)\|f - g\| \qquad (2.10)$$

holds for every $g \in C$.

Proof Let us fix a pointwise Lipschitzian mapping T, and the function $\alpha(\cdot)$ satisfying (2.10). Fix also $c \in [0, 1]$ and $x, y \in C$. Let $\delta : [0, 2] \to [0, 1]$ denote the modulus of convexity for X. From the definition of the modulus of convexity it follows easily that

$$2\min(c, 1-c)\delta(\|u - v\|) \leq 1 - \|cu + (1-c)v\| \qquad (2.11)$$

for $\|u\|, \|v\| \leq 1$. Define a function

$$d(t) = \frac{1}{2}\int_0^t \delta(s)ds \; t \in [0, 2],$$

and extend it to the interval $(2, \infty)$ by $d(t) = d(2) + \frac{1}{2}\delta(2)(t - 2)$. The function d is straightforwardly shown to be strictly increasing, continuous, and convex, with the property that $d(0) = 0$. Moreover, it satisfies the inequality

$$d(t) \leq \delta(t), \; t \in [0, 2]. \qquad (2.12)$$

Additionally, the function defined by

$$s \mapsto \frac{d(s)}{s} \qquad (2.13)$$

is an increasing function on the interval $[0, 2]$. By employing inequality (2.12) and the fact that $c(1 - c) \leq \min(c, 1 - c)$, we can deduce from (2.11) that

$$2c(1 - c)d(\|u - v\|) \leq 1 - \|cu + (1-c)v\|. \qquad (2.14)$$

Set $w = cx + (1 - c)y$ and

2.2 Common Fixed-Point Existence for Asymptotic Pointwise Nonexpansive Semigroups

$$u = \frac{T(y) - T(w)}{c\|x - y\|\alpha(w)}, \quad v = \frac{T(w) - T(x)}{(1-c)\|x - y\|\alpha(w)}. \tag{2.15}$$

By applying the definition of pointwise Lipschitzian mappings, it can be readily demonstrated that $\|u\|, \|v\| \leq 1$. It follows from the definitions of u and v that

$$u - v = \frac{cT(x) + (1-c)T(y) - T(w)}{c(1-c)\alpha(w)\|x - y\|} \tag{2.16}$$

$$cu + (1-c)v = \frac{T(x) - T(y)}{\alpha(w)\|x - y\|}. \tag{2.17}$$

By substituting (2.16) and (2.17) into (2.14), and multiplying both sides by $\|x - y\|$, we obtain the inequality

$$2c(1-c)\|x - y\|d\left(\frac{\|cT(x) + (1-c)T(y) - T(w)\|}{c(1-c)\alpha(w)\|x - y\|}\right)$$
$$\leq \|x - y\| - \frac{\|T(x) - T(y)\|}{\alpha(w)}. \tag{2.18}$$

By applying the inequality

$$c(1-c)\|x - y\| \leq \frac{\mathrm{diam}(C)}{4},$$

alongside the fact that the function defined in (2.13) is increasing, we can derive from (2.18) that

$$\frac{1}{2}\mathrm{diam}(C)d\left(\frac{4\|T(w) - cT(x) - (1-c)T(y)\|}{\mathrm{diam}(C)\alpha(w)}\right)$$
$$\leq \|x - y\| - \frac{\|T(x) - T(y)\|}{\alpha(w)}. \tag{2.19}$$

We are now prepared to define the function γ_2 as follows:

$$\gamma_2(t) = \frac{1}{2}\mathrm{diam}(C)d\left(\frac{4t}{\mathrm{diam}(C)}\right).$$

Upon substituting γ_2 into (2.19), we arrive at (2.9), thereby concluding the proof. □

It is important to note that the function γ_2, as defined in Lemma 2.1, depends only on $diam(C)$. Therefore, it satisfies the inequality (2.9) for all pointwise Lipschitzian mappings acting within C.

The following parallelogram inequality due to Xu [106, Theorem 2] (see also [50, Proposition 3.4]) provides an important characterisation of uniformly convex Banach spaces.

Proposition 2.1 *A Banach space X is uniformly convex if and only if for every $d > 1$ there exists a continuous, strictly increasing and convex function $\lambda : [0, \infty) \to [0, \infty)$, $\lambda(t) = 0$, such that*

$$\|cw + (1-c)v\|^2 \leq c\|w\|^2 + (1-c)\|v\|^2 - c(1-c)\lambda(\|w - v\|) \qquad (2.20)$$

holds for any $c \in [0, 1]$ and all $w, v \in X$ such that $\|w\| \leq d$ and $\|v\| \leq d$.

The parallelogram inequality (2.20) provides a crucial element for proving the existence of a common fixed point.

Theorem 2.3 *Assume that X is uniformly convex. Let $\mathcal{T}$ be an asymptotic pointwise nonexpansive semigroup on a nonempty, bounded, closed, and convex set $C \subset X$. Under these conditions, $\mathcal{T}$ possesses a common fixed point, and the set of common fixed points, denoted $F(\mathcal{T})$, is closed and convex.*

Proof Fix an arbitrary $x \in C$. Let us define the type function φ by the formula

$$\varphi(y) = \limsup_{t \to \infty} \|T_t(x) - y\|^2.$$

By Remark 2.2, there exists an element $z \in C$ such that $\varphi(z) = \min\{\varphi(y) : y \in C\}$. We will prove now that $T_t(z)$ satisfies the following Cauchy type condition: to every $\epsilon > 0$ there exists a $t_\epsilon \in J$ such that

$$\|T_s(z) - T_u(z)\| \leq \epsilon$$

for all $s, u \in J$ such that $s, u \geq t_\epsilon$. To this end, let us fix any $t, s, u \in J$ and any $\epsilon > 0$. By applying inequality (2.20) from Proposition 2.1 with $w = T_{s+u+t}(x) - T_s(z)$, $v = T_{s+u+t}(x) - T_u(z)$, $d = diam(C) + \|x\|$ and $c = \frac{1}{2}$, we get

$$\|T_{s+u+t}(x) - \frac{1}{2}(T_s(z) + T_u(z))\|^2 \qquad (2.21)$$
$$\leq \frac{1}{2}\|T_{s+u+t}(x) - T_s(z)\|^2 + \frac{1}{2}\|T_{s+u+t}(x) - T_u(z)\|^2 - \frac{1}{4}\lambda(\|T_s(z) - T_u(z)\|).$$

Taking the limit superior of both sides of (2.21) as $t \to \infty$ and utilising the definition of φ along with the pointwise Lipschitzian properties of $\mathcal{T}$, we arrive at the inequality

$$\varphi\left(\frac{T_s(z) + T_u(z)}{2}\right) \leq \frac{1}{2}\left(\alpha_s(z)^2 + \alpha_u(z)^2\right)\varphi(z) - \frac{1}{4}\lambda(\|T_s(z) - T_u(z)\|). \qquad (2.22)$$

Leveraging the convexity of C and the minimality property of φ at z, we can conclude that

$$\varphi(z) \leq \varphi\left(\frac{T_s(z) + T_u(z)}{2}\right)$$

for all $s, u \in J$. Therefore, by (2.22),

2.2 Common Fixed-Point Existence for Asymptotic Pointwise Nonexpansive Semigroups

$$\varphi(z) \le \frac{1}{2}\big(\alpha_s(z)^2 + \alpha_u(z)^2\big)\varphi(z) - \frac{1}{4}\lambda(\|T_s(z) - T_u(z)\|),$$

which implies that

$$\lambda(\|T_s(z) - T_u(z)\|) \le 2\big(\alpha_s(z)^2 + \alpha_u(z)^2\big)\varphi(z) - 4\varphi(z). \tag{2.23}$$

It follows from (2.23) that

$$\lim_{s,u \to \infty} \lambda(\|T_s(z) - T_u(z)\|) = 0. \tag{2.24}$$

Recalling that λ is a continuous function satisfying $\lambda(t) = 0$ if and only if $t = 0$, we can infer from (2.24) that $\lim_{s,u \to \infty} \|T_s(z) - T_u(z)\| = 0$. Thus, we can choose a $t_\epsilon \in J$ such that $\|T_s(z) - T_u(z)\| \le \epsilon$ for all $s, u \in J$ with $s, u \ge t_\epsilon$, as asserted. Given the completeness of X and closedness of C, there exists then an element $f \in C$ such that

$$\lim_{s \to \infty} \|f - T_s(z)\| = 0. \tag{2.25}$$

We now claim that $f \in F(\mathcal{F})$. To demonstrate this, consider any $t, s \in J$ and observe that

$$\|T_t(f) - T_{t+s}(z)\| \le \alpha_t(f)\|f - T_s(z)\|.$$

Taking the limit as $s \to \infty$ and employing Eq. (2.25), we obtain $\|T_t(f) - f\| \le 0$ for every $t \in J$. This implies that $f \in F(T_t)$ for every $t \in J$. Consequently, we conclude that $f \in F(\mathcal{T})$, which means that f is a common fixed point.

To prove that $F(\mathcal{T})$ is closed, it suffices to demonstrate that every $F(T_t)$ is closed. To this end, fix $t \in J$, let $v_n \in F(T_t)$ for every $n \in \mathbb{N}$, and let $v_n \to v$. Observe that

$$\|T_t(v) - v\| \le \|T_t(v) - v_n\| + \|v_n - v\| = \|T_t(v) - T_t(v_n)\| + \|v_n - v\|$$
$$\le \alpha_t(v)\|v - v_n\| + \|v_n - v\| \to 0,$$

as $n \to \infty$. Thus, $v \in F(T_t)$, which proves that $F(\mathcal{T})$ is closed.

It only remains to show that $F(\mathcal{T})$ is convex. Let $f_1, f_2 \in F(\mathcal{T})$ and let $c \in [0, 1]$. We need to demonstrate that $f = cf_1 + (1-c)f_2 \in F(\mathcal{T})$. Fix any $t \in J$. It follows from Lemma 2.1 that

$$\|T_t(f) - f\| = \|T_t(f) - cf_1 - (1-c)f_2\| = \|T_t(f) - cT_t(f_1) - (1-c)T_t(f_2)\|$$
$$\le \alpha_t(f)\gamma_2^{-1}\bigg(\|f_1 - f_2\| - \frac{\|T_t(f_1) - T_t(f_2)\|}{\alpha_t(f)}\bigg)$$
$$= \alpha_t(f)\gamma_2^{-1}\bigg(\|f_1 - f_2\| - \frac{\|f_1 - f_2\|}{\alpha_t(f)}\bigg).$$

Since γ_2^{-1} is a continuous function that equals zero only at zero, and given that $\alpha_t(f) \to 1$ as $t \to \infty$, we conclude that

$$\|T_t(f) - f\| \to 0. \tag{2.26}$$

Take any $s, t \in J$ and note that, by (2.26), the right-hand side of the inequality

$$\|T_s(f) - f\| \leq \|T_s(f) - T_{t+s}(f)\| + \|T_{t+s}(f) - f\| \leq \alpha_s(f)\|f - T_t(f)\| + \|T_{t+s}(f) - f\|$$

tends to zero as $t \to \infty$. This implies that $f \in F(T_s)$ for any $s \in J$. Thus, f is a common fixed point for the semigroup $\mathcal{T}$.

The proof of Theorem 2.3 is complete. □

Remark 2.4 Assume that X and C are defined as in Theorem 2.3. Let $T : C \to C$ be a nonexpansive mapping, and let $\mathcal{T}$ denote a semigroup of all iterations of the form $\{T^k : k = 0, 1, 2, 3, ...\}$ of T. As noted in Remark 1.1, we have $F(\mathcal{T}) = F(T)$. Consequently, Theorem 2.3 guarantees the existence of a fixed point for T. Thus, Theorem 2.3 serves as a direct generalisation of Browder's fixed point theorem for a single nonexpansive mapping.

By combining Theorem 2.3 with Proposition 1.4, we immediately obtain the following fixed point result.

Theorem 2.4 *Assume X is uniformly convex. Let $\mathcal{T}$ be a pointwise Lipschitzian semigroup on C. Assume that all mappings $T_t \in \mathcal{T}$ are continuously Fréchet differentiable on an open convex set A containing C. Suppose that for each $x \in C$,*

$$\limsup_{t \to \infty} \|(T_t)'_x\| \leq 1. \tag{2.27}$$

Then, $\mathcal{T}$ has a common fixed point in C and the set $F(\mathcal{T})$ of common fixed points is closed convex.

By applying Theorem 2.4 in the context of Example 1.7, we can immediately infer that the operator T must possess a fixed point within the unit ball of the Hilbert space l^2. It is important to note that this conclusion would not have been attainable with only the nonexpansive version of the existence theorem at our disposal.

2.3 Notes

The existence of common fixed points for families of contractions and nonexpansive mappings has been a subject of investigation since the early 1960s, as referenced in DeMarr [33], Browder [17], and Belluce and Kirk [13, 14]. Moreover, Lim [81] and Bruck [18, 19] have contributed significantly to this field. The asymptotic approach for identifying common fixed points of semigroups of

2.3 Notes

Lipschitzian mappings (but not pointwise Lipschitzian mappings) has also been explored for some time; see, for example, Tan and Xu [105]. It is noteworthy to mention studies focusing on the special case where the parameter set for the semigroup corresponds to $\{0, 1, 2, 3, ...\}$ and $T_n = T^n$, representing the n-th iterate of an asymptotic pointwise nonexpansive mapping. Specifically, this refers to a mapping $T : C \to C$ for which there exists a sequence of functions $\alpha_n : C \to [0, \infty)$ satisfying $\|T^n(x) - T^n(y)\| \leq \alpha_n(x)\|x - y\|$ and $\limsup_{n \to \infty} \alpha_n(x) \leq 1$. Kirk and Xu [50] established the existence of fixed points for asymptotic pointwise contractions and asymptotic pointwise nonexpansive mappings within Banach spaces. Furthermore, the existence of common fixed points for asymptotic pointwise nonexpansive semigroups was demonstrated by Kozlowski [59]. The content of this chapter is primarily based on the latter work.

Chapter 3
Our Toolset for Constructing Common Fixed Points

Abstract As suggested by its title, this chapter introduces several key concepts and technical results essential for proving the convergence of the common fixed point construction processes discussed in the subsequent chapters. Among other things, we present a version of the Demiclosedness Principle adapted to our pointwise Lipschitzian semigroup context, specifically for the case when X is a uniformly convex Banach space with the Opial property, as well as its counterpart for spaces that are both uniformly convex and uniformly smooth.

3.1 Preliminaries

The following elementary lemmas will be used in this book.

Lemma 3.1 ([20]) *Suppose $\{r_k\}$ is a bounded sequence of real numbers and $\{d_{k,n}\}$ is a doubly-index sequence of real numbers which satisfy*

$$\limsup_{k \to \infty} \limsup_{n \to \infty} d_{k,n} \leq 0, \text{ and } r_{k+n} \leq r_k + d_{k,n}$$

for each $k, n \geq 1$. Then $\{r_k\}$ converges to an $r \in \mathbb{R}$.

Lemma 3.2 ([100, Lemma 1]) *Let $\{t_n\}$ be a sequence of real numbers and let $\tau \in \mathbb{R}$ be such that*

$$\liminf_{n \to \infty} t_n \leq \tau \leq \limsup_{n \to \infty} t_n$$

and

$$\lim_{n \to \infty} (t_{n+1} - t_n) = 0.$$

Then, τ is a cluster point of the sequence $\{t_n\}$.

The concept of bounded away sequences of real numbers will be extensively utilised throughout the following chapters.

Definition 3.1 A sequence $\{t_n\}$ within the interval $(0, 1)$ is said to be bounded away from 0 if there exists a constant $a \in (0, 1)$ such that $t_n > a$ for all $n \in \mathbb{N}$. Similarly, $\{t_n\}$ is called bounded away from 1 if there exists $b \in (0, 1)$ such that $t_n < b$ for all $n \in \mathbb{N}$.

The following property of uniformly convex Banach spaces will play an important role in our convergence results.

Lemma 3.3 ([98, 114]) *Let X be a uniformly convex Banach space and let a sequence $\{c_n\}$ within the interval $(0, 1)$ be bounded away from 0 and 1. Let the sequences $\{u_n\}, \{v_n\}$ of elements of X be such that*

$$\limsup_{n \to \infty} \|u_n\| \leq a, \ \limsup_{n \to \infty} \|v_n\| \leq a, \ \lim_{n \to \infty} \|c_n u_n + (1 - c_n) v_n\| = a.$$

Then, $\lim_{n \to \infty} \|u_n - v_n\| = 0$.

Let us recall the definition of uniformly smooth Banach spaces.

Definition 3.2 A Banach space X is said to be uniformly smooth if and only if the limit

$$\lim_{t \to 0} \frac{\|x + ty\| - \|x\|}{t}$$

exists uniformly for all $x, y \in S_X$, where S_X denotes the unit sphere of X.

It is established that if X is uniformly smooth, then the following subdifferential inequality holds, as referenced in [103, 105].

Proposition 3.1 *Let X be a uniformly smooth Banach space. Then,*

$$\frac{1}{2}\|x\|^2 + \langle h, J(x) \rangle \leq \frac{1}{2}\|x + h\|^2 \leq \frac{1}{2}\|x\|^2 + \langle h, J(x) \rangle + \|h\|^2 \qquad (3.1)$$

for all $x, h \in X$, where J is the normalised duality map from X to X^* defined by

$$J(x) = \{x^* \in X^*; \langle x, x^* \rangle = \|x\|^2 = \|x^*\|^2\}.$$

Note that $\langle \cdot, \cdot \rangle$ represents the duality pairing between X and X^*.

Let $\mathcal{T}$ be a pointwise Lipschitzian semigroup on C. By defining $a_0 \equiv 1$ and $a_t(x) = \max(\alpha_t(x), 1)$ for $t > 0$, we can assert, without loss of generality, that the semigroup $\mathcal{T}$ is asymptotic pointwise nonexpansive if the following conditions hold for every $x, y \in C$ and for all $t \in J$,

$$\|T_t(x) - T_t(y)\| \leq a_t(x)\|x - y\|$$

where $a_t(x) \geq 1$ for every $x \in C$, $t \in J$, and $\lim_{t \to \infty} a_t(x) = 1$. Define $b_t(x) = a_t(x) - 1$ and note that $\lim_{t \to \infty} b_t(x) = 0$. These conventions will be consistently applied throughout the rest of this book.

3.1 Preliminaries

Remark 3.1 Assume that for a given sequence $\{t_i\}$ of elements from J, the series of the remainders of $a_{t_i}(x)$ converges, which means that

$$\sum_{i=1}^{+\infty} b_{t_i}(x) < \infty. \tag{3.2}$$

It follows from the elementary inequality

$$1 + \sum_{i=k}^{N} b_{t_i}(x) \leq \prod_{i=k}^{N} \left(1 + b_{t_i}(x)\right) \leq \exp\left(\sum_{i=k}^{N} b_{t_i}(x)\right) \tag{3.3}$$

that the infinite product $\prod_{i=1}^{+\infty} a_{t_i}(x)$ also converges. In addition, it can be readily derived from (3.3) that, given the assumption (3.2), we have

$$\lim_{k \to \infty} \prod_{i=k}^{+\infty} a_{t_i}(x) = 1.$$

These facts will be utilised frequently in our discussions of the fixed point construction processes.

The concept of a generating set for J is commonly employed in this book to streamline calculations.

Definition 3.3 Let $\mathcal{T} = \{T_t\}_{t \in J}$ be a pointwise Lipschitzian semigroup on C. A set $A \subset J$ is called a generating set for the parameter semigroup J if for every $0 < u \in J$ there exist $m \in \mathbb{N}$, $s \in A$, $t \in A$ such that $u = ms + t$. Consequently, if $T_s(w) = w$ for every $s \in A$, where A is generating for J, then $w \in F(\mathcal{T})$.

The above-mentioned definition of a generating set is often utilised for parameter semigroups J underlying semigroups of nonlinear operators. This usage is particularly advantageous as it streamlines calculations and proofs in two key scenarios that are critical from the perspectives of fixed point theory and its applications: when $J = [0, +\infty)$ and when J is a subgroup of the additive semigroup $\mathbb{N}_0 := \mathbb{N} \cup \{0\}$. Note that any interval $A = [0, \beta)$ with $\beta > 0$, can act as a generating set for $J = [0, +\infty)$. On the other hand, the set $A = \{\alpha\}$ generates $J = \{n\alpha : n = 0, 1, 2, 3, ...\}$. In particular, the set $A = \{1\}$ generates the entire semigroup $\mathbb{N}_0$.

Remark 3.2 In the theory of additive semigroups of nonnegative integers, an alternative definition of a set of generators is commonly used. Given a subsemigroup J of $\mathbb{N}_0$, we say a subset $G \subset J$ is a set of generators of J if every number in J can be expressed as the sum of nonnegative multiples of elements in G. We say J is finitely generated if such a subset G exists and is finite. It is straightforward to see that, in this case, if $T_s(w) = w$ for every $s \in G$, where G is a set of generators of J, then $w \in F(\mathcal{T})$. Furthermore, it is well known that any subsemigroup J of $\mathbb{N}_0$ is finitely generated. These two facts will be very useful when considering common fixed point construction processes for semigroups of nonlinear operators parameterised by integers, as they allow us to reduce the problem of finding a common fixed point for the entire semigroup to that of seeking a common fixed point for a finite number of operators.

By utilising the concept of a generating set, we can also demonstrate a particularly useful result for the case when $J = [0, +\infty)$. This result indicates that, in certain situations, instead of seeking a common fixed point for the entire semigroup, we can focus on finding a fixed point shared by just two mappings, or even just one mapping, which is often a significantly simpler task.

Proposition 3.2 *Let $\mathcal{T}$ be a continuous semigroup of operators acting within C, where C is a nonempty, bounded, closed, and convex subset of a Banach space X. Assume that $\mathcal{T}$ is parametrised by $J = [0, +\infty)$. Suppose that there exist two numbers $\alpha, \beta \in J$ such that the ratio $\frac{\alpha}{\beta}$ is an irrational number. Then,*

$$F(\mathcal{T}) = F(T_\alpha) \cap F(T_\beta).$$

Proof Since

$$F(\mathcal{T}) = \bigcap_{t \in J} F(T_t) \subset F(T_\alpha) \cap F(T_\beta),$$

it suffices to prove that if $x \in F(T_\alpha) \cap F(T_\beta)$ then $x \in F(\mathcal{T})$. Let us fix such an $x \in F(T_\alpha) \cap F(T_\beta)$. For any $r \in \mathbb{R}$, denote by $\lfloor r \rfloor$ the largest integer less than or equal to r. We know that the set A_n defined by the formula

$$A = \left\{ n\frac{\alpha}{\beta} - \left\lfloor n\frac{\alpha}{\beta} \right\rfloor : n \in \mathbb{N}_0 \right\}$$

is dense in $[0, 1)$. Denote $m_n = \lfloor n\frac{\alpha}{\beta} \rfloor$ and $\gamma_n = n\frac{\alpha}{\beta} - m_n$. Using this notation, we have

$$\beta \gamma_n + m_n \beta = n\alpha.$$

Consequently,

$$T_{\beta \gamma_n}(T_{m_n \beta}(x)) = T_{\beta \gamma_n + m_n \beta}(x) = T_{n\alpha}(x) = x,$$

and, since $T_{m_n \beta}(x) = x$, we conclude that $x \in F(T_{\beta \gamma_n})$ for every $n \in \mathbb{N}_0$. This implies that $x \in F(T_s)$ for every $s \in \beta A$, which is a dense subset of $[0, \beta)$. Take any $t \in [0, \beta)$. There exists then a sequence $\{t_n\}$ of elements of the set βA such that $t_n \to t$. We conclude that

$$\|T_t(x) - x\| \leq \|T_t(x) - T_{t_n}(x)\| + \|T_{t_n}(x) - x\| = \|T_t(x) - T_{t_n}(x)\| \to 0,$$

by the continuity of the semigroup $\mathcal{T}$. Therefore, $x \in F(T_t)$ for every $t \in [0, \beta)$. Finally, $x \in F(T_t)$ for every $t \in [0, +\infty)$ because $[0, \beta)$ is a generating set for $[0, +\infty)$. □

Definition 3.4 Let $\mathcal{T}$ be a pointwise Lipschitzian semigroup on C. A sequence $\{x_k\}_{k\in\mathbb{N}}$ of elements in C is referred to as a regular sequence with respect to the semigroup $\mathcal{T}$—or simply as a regular sequence, when the context is clear—if for every $t \in J$,
$$M_t := \sup\{a_t(x_k) : k \in \mathbb{N}\} < \infty.$$

Note that M_t depends only on $t \in J$ and the sequence $\{x_k\}_{k\in\mathbb{N}}$, not on any individual term of this sequence.

Observe that if every $a_t(\cdot)$ is a bounded function on C, then every sequence is regular with respect to $\mathcal{T}$. This is, for instance, the case when $\mathcal{T}$ is a Lipschitzian semigroup, meaning that each a_t is a constant, and in particular when $\mathcal{T}$ is a nonexpansive semigroup.

3.2 Approximate Fixed-Point Sequences

The technique of approximate fixed point sequences will be essential in demonstrating convergence to common fixed points for semigroups of mappings. Recall that for a mapping $T : C \to C$, a sequence $\{x_k\}$ of elements in C is termed an approximate fixed point sequence for T if $\|T(x_k) - x_k\| \to 0$ as $k \to \infty$.

We start with the following technical result.

Lemma 3.4 *Let C be a nonempty, bounded, closed, and convex subset of a Banach space X. Let $\mathcal{T} \in \mathcal{APNS}(C)$. Assume that a sequence $\{x_k\}_{k\in\mathbb{N}}$ of elements in C is a regular sequence with respect to the semigroup $\mathcal{T}$. If the sequence $\{x_k\}$ is an approximate fixed point sequence for T_s for an $s \in J$ then $\{x_k\}$ is an approximate fixed point sequence for T_{ms} for any $m \in \mathbb{N}$, i.e., $\|T_{ms}(x_n) - x_n\| \to 0$ as $n \to \infty$.*

Proof Fix an $m \in \mathbb{N}$. It follows from the regularity of $\{x_k\}$ that there exists a finite constant $M > 0$ such that

$$\sum_{j=1}^{m-1} \sup\{a_{js}(x_k) : k \in \mathbb{N}\} \leq M.$$

It follows from

$$\|T_{ms}(x_n) - x_n\| \leq \sum_{j=1}^{m-1} \|T_{(j+1)s}(x_n) - T_{js}(x_n)\| + \|T_s(x_n) - x_n\|$$

$$\leq \|T_s(x_n) - x_n\|\left(\sum_{j=1}^{m-1} a_{js}(x_n) + 1\right) \leq (M+1)\|T_s(x_n) - x_n\|$$

that

$$\lim_{n\to\infty} \|T_{ms}(x_n) - x_n\| = 0,$$

which completes the proof. □

The previous lemma provides a critical component for the proof of our next result, which shows that in the case of a regular sequence $\{x_k\}$, it suffices to prove that it is fixed point approximating at every parameter p from the set generating J in order to ensure that it is fixed point approximating at every $q \in J$.

Theorem 3.1 *Let C be a nonempty, bounded, closed, and convex subset of a Banach space X. Let $\mathcal{T} \in \mathcal{APNS}(C)$. If a regular sequence $\{x_n\}$ is an approximate fixed point sequence for $T_p \in \mathcal{T}$ for any $p \in A$ where A is a generating set for J then $\{x_n\}$ is an approximate fixed point sequence for T_q for any $q \in J$.*

Proof Let $q = ms + t$, where $s, t \in A$ and $m \in \mathbb{N}$. In order to prove that

$$\lim_{n\to\infty} \|T_{ms+t}(x_n) - x_n\| = 0,$$

consider the following series of inequalities,

$$\begin{aligned}
\|T_{ms+t}(x_n) - x_n\| &\leq \|T_{ms+t}(x_n) - T_{ms}(x_n)\| + \|T_{ms}(x_n) - x_n\| \\
&\leq a_{ms}(x_n)\|T_t(x_n) - x_n\| + \|T_{ms}(x_n) - x_n\| \\
&\leq \sup\{a_{ms}(x_k) : k \in \mathbb{N}\}\|T_t(x_n) - x_n\| + \|T_{ms}(x_n) - x_n\|.
\end{aligned}$$

The right-hand side tends to zero because $\{x_k\}$ is a regular approximate sequence for T_t, and because $\|T_{ms}(x_n) - x_n\| \to 0$ by Lemma 3.4. □

For the important case when $J = [0, +\infty)$, assuming equicontinuity at zero of $\mathcal{T}$, it actually suffices to prove that a regular sequence is an approximate fixed point sequence at a dense subset of a generating set $[0, 1)$, as demonstrated below.

Theorem 3.2 *Let $\mathcal{T} \in \mathcal{APNS}(C)$ be an equicontinuous semigroup with $J = [0, +\infty)$, which implies that $A = [0, 1)$ is generating for J. Assume that B is a dense subset of A. If a regular sequence $\{x_k\}$ is an approximate fixed point sequence for T_t for every $t \in B$ then $\{x_k\}$ is an approximate fixed point sequence for T_s for every $s \in A$.*

Proof Fix an $s \in A$. There exists then a sequence $\{s_n\}$ of elements in B such that $s_n \to s^+$. Fix temporarily an $n \in \mathbb{N}$ and observe that for any $k \in \mathbb{N}$,

$$\|T_{s_n}(x_k) - T_s(x_k)\| \leq a_s(x_k)\|T_{s_n-s}(x_k) - x_k\| \leq M_s\|T_{s_n-s}(x_k) - x_k\|, \quad (3.4)$$

where $M_s = \sup\{a_s(x_p) : p \in \mathbb{N}\}$. Note that M_s is finite because $\{x_k\}$ is regular with respect to $\mathcal{T}$. Fix $\varepsilon > 0$. Since $\mathcal{T}$ is equicontinuous, it follows that there exists $n_0 \in \mathbb{N}$ such that

$$\sup_{x \in C} \|T_{s_n - s}(x) - x\| < \frac{\varepsilon}{2M_s} \tag{3.5}$$

holds for all $n \geq n_0$. In particular, by taking $n = n_0$ and substituting (3.5) into the inequality (3.4) we conclude that the inequality

$$\|T_{s_{n_0}}(x_k) - T_s(x_k)\| \leq \frac{\varepsilon}{2} \tag{3.6}$$

is satisfied for every $k \in \mathbb{N}$. Since $\{x_k\}$ is an approximate fixed point for $T_{s_{n_0}}$, we can find $k_0 \in \mathbb{N}$ such that, for every natural $k \geq k_0$,

$$\|T_{s_{n_0}}(x_k) - x_k\| < \frac{\varepsilon}{2}. \tag{3.7}$$

Observe that, based on inequalities (3.6) and (3.7),

$$\|T_s(x_k) - x_k\| \leq \|T_s(x_k) - T_{s_{n_0}}(x_k)\| + \|T_{s_{n_0}}(x_k) - x_k\| \leq \varepsilon,$$

holds for any $k \geq k_0$. Thus, $\{x_k\}$ is an approximate fixed point for T_s, as claimed. □

3.3 The Demiclosedness Principle for Asymptotic Pointwise Nonexpansive Semigroups

The technique commonly known as the Demiclosedness Principle offers a valuable method for proving convergence in the weak topology of fixed point construction processes. This approach involves specifying certain conditions on a Banach space X that ensure an element of X is a fixed point if it is the weak limit of an approximate point sequence. There are multiple versions of the Demiclosedness Principle applicable to asymptotic nonexpansive mappings and their semigroups. For further details, see works by Li and Sims [80], Gornicki [43], Xu [107], Kozlowski [61, 62], Kozlowski and Sims [77]. In this section, we will prove two versions of this Principle adapted to our semigroup setting. The following general result will play a crucial role in their proofs.

Theorem 3.3 *Let C be a nonempty, bounded, closed, and convex subset of a uniformly convex Banach space X. Let $\mathcal{T} \in \mathcal{APNS}(C)$. Assume that a regular sequence $\{x_n\}$ of elements of C is an approximate fixed point sequence for an operator $T_s \in \mathcal{T}$. Let φ be a type defined for every $x \in C$ by,*

$$\varphi(x) = \limsup_{n \to \infty} \|x_n - x\|. \tag{3.8}$$

If φ attains its minimum at $w \in C$, then $w \in F(T_{ks})$ for any natural k.

Proof Let us fix two natural numbers: $n \geq 1$ and $m > 2$. Observe that

$$\|T_{ms}(x_n) - x\| \leq \sum_{i=1}^{m} \|T_{is}(x_n) - T_{(i-1)s}(x_n)\| + \|x_n - x\|$$

$$\leq \|T_s(x_n) - x_n\| \Big(\sum_{i=2}^{m} a_{(i-1)s}(x_n) + 1\Big) + \|x_n - x\|. \quad (3.9)$$

Since the sequence $\{x_n\}$ is regular with respect to $\mathcal{T}$, it follows that there exists a finite constant K_m, depending only on m, such that

$$\sum_{i=2}^{m} a_{(i-1)s}(x_n) \leq K_m.$$

By substituting this constant into (3.9) we obtain the inequality

$$\|T_{ms}(x_n) - x\| \leq (K_m + 1)\|T_s(x_n) - x_n\| + \|x_n - x\|.$$

By taking $\limsup_{n \to \infty}$ of both sides, and using the fact that $\{x_n\}$ is an approximate fixed point sequence for T_s, we conclude that

$$\limsup_{n \to \infty} \|T_{ms}(x_n) - x\| \leq \limsup_{n \to \infty} \|x_n - x\| = \varphi(x). \quad (3.10)$$

On the other hand, it follows from Lemma 3.4 that $\|T_{ms}(x_n) - x_n\| \to 0$ as $n \to \infty$. Thus,

$$\varphi(x) \leq \limsup_{n \to \infty} \|x_n - T_{ms}(x_n)\| + \limsup_{n \to \infty} \|T_{ms}(x_n) - x\| = \limsup_{n \to \infty} \|T_{ms}(x_n) - x\|,$$

which together with (3.10) implies that

$$\varphi(x) = \limsup_{n \to \infty} \|T_{ms}(x_n) - x\|. \quad (3.11)$$

By utilising Eq. (3.11) along with the property that T_{ms} is asymptotic pointwise nonexpansive, we derive the inequality

$$\varphi(T_{ms}(x)) = \limsup_{n \to \infty} \|T_{ms}(x_n) - T_{ms}(x)\| \leq a_{ms}(x) \limsup_{n \to \infty} \|x_n - x\| = a_{ms}(x)\varphi(x).$$

Apply this inequality to $x = w$, keeping in mind that $\lim_{m \to \infty} a_{ms}(w) = 1$, and take the limits of both sides as $m \to \infty$ to obtain the inequality

$$\lim_{m \to \infty} \varphi(T_{ms}(w)) \leq \varphi(w). \quad (3.12)$$

3.3 The Demiclosedness Principle for Asymptotic Pointwise Nonexpansive Semigroups

By assumption, $\varphi(w) = \inf\{\varphi(x) : x \in C\}$, which, together with (3.12) implies that

$$\lim_{m \to \infty} \varphi(T_{ms}(w)) = \varphi(w). \tag{3.13}$$

Choose $d > 0$ such that C is contained within a ball centred at zero and radius d. By Proposition (2.1), applied with this $d > 0$, there exists a continuous function $\lambda : [0, \infty) \to [0, \infty)$ such that $\lambda(t) = 0$ if and only $t = 0$, and the following parallelogram inequality holds for any $\alpha \in [0, 1]$ and all $x, y \in C$,

$$\|\alpha x + (1 - \alpha) y\|^2 \leq \alpha \|x\|^2 + (1 - \alpha) \|y\|^2 - \alpha(1 - \alpha) \lambda(\|x - y\|). \tag{3.14}$$

Applying (3.14) to $x = x_n - w$, $y = x_n - T_{ms}(w)$ and $\alpha = \frac{1}{2}$ we obtain the inequality

$$\left\| x_n - \frac{1}{2}(w + T_{ms}(w)) \right\|^2 \leq \frac{1}{2}\|x_n - w\|^2 + \frac{1}{2}\|x_n - T_{ms}(w)\|^2 - \frac{1}{4}\lambda(\|T_{ms}(w) - w\|).$$

Applying to both sides $\limsup_{n \to \infty}$ and recalling that $\varphi(w) = \inf\{\varphi(x) : x \in C\}$, we obtain

$$\varphi(w)^2 \leq \frac{1}{2}\varphi(w)^2 + \frac{1}{2}\varphi(T_{ms}(w))^2 - \frac{1}{4}\lambda(\|T_{ms}(w) - w\|),$$

which implies that

$$\lambda(\|T_{ms}(w) - w\|) \leq 2\varphi(T_{ms}(w))^2 - 2\varphi(w)^2. \tag{3.15}$$

By taking the limits of both sides of (3.15) as $m \to \infty$ and using the equality (3.13), we conclude that

$$\lim_{m \to \infty} \lambda(\|T_{ms}(w) - w\|) = 0,$$

which, by the properties of λ, implies that $T_{ms}(w) \to w$ as $m \to \infty$. Using the continuity of T_{ks}, we infer from this that

$$T_{ks}(T_{ms}(w)) \to T_{ks}(w), \tag{3.16}$$

when $m \to \infty$. Fix any natural number k. Using the same argument as above, we can conclude that $T_{(k+m)s}(w) \to w$ as $m \to \infty$. Thus,

$$T_{ks}(T_{ms}(w)) = T_{(k+m)s}(w) \to w. \tag{3.17}$$

Given the uniqueness of the limit, the Eqs. (3.16) and (3.17) establish that $T_{ks}(w) = w$, as claimed. □

Remark 3.3 It is worthwhile to mention a specific, and very classical, case of the semigroup $\mathcal{T} = \{T^k : k = 0, 1, ...\}$, where $T : C \to C$ is nonexpansive. As we noted in Remark 1.1, in this case $F(\mathcal{T}) = F(T)$, which is—by Browder's Theorem—nonempty. Thus, Theorem 3.3, when applied to this case with $T_s = T$ and $k = 1$, indicates that if φ, defined by an approximate fixed point sequence as in (3.8), attains its minimum at $w \in C$, then w is a fixed point for operator T.

Theorem 3.3 and its proof are not only very elegant but also powerful. This result simplifies proofs of any versions of the Demiclosedness Principle by requiring only that the type φ, defined by a regular approximate fixed point sequence $\{x_n\}$, attains its infimum at an element $w \in C$, which is a weak-limit of $\{x_n\}$. To achieve this, we will first consider the case where the Banach space X possess the Opial property.

Definition 3.5 ([87]) A Banach space X is said to have the Opial property if, for each sequence $\{x_n\}$ of elements from X that weakly converges to a point $x \in X$ (denoted as $x_n \rightharpoonup x$), and for any $y \in X$ such that $y \neq x$, the following two equivalent inequalities hold,

$$\liminf_{n \to \infty} \|x_n - x\| < \liminf_{n \to \infty} \|x_n - y\|,$$

$$\limsup_{n \to \infty} \|x_n - x\| < \limsup_{n \to \infty} \|x_n - y\|.$$

Theorem 3.4 (The Demiclosedness Principle 1) *Let X be a uniformly convex Banach space with the Opial property. Let C be a nonempty, bounded, closed, and convex subset of X, and let $\mathcal{T} \in \mathcal{APNS}(C)$. Assume that there exists $w \in X$ and a regular sequence $\{x_n\}$ of elements of C such that $x_n \rightharpoonup w$. If $\{x_n\}$ is an approximate fixed point sequence for an operator $T_s \in \mathcal{T}$, then $w \in F(T_{ks})$ for any natural k.*

Proof Since $x_n \rightharpoonup w$, by the Opial property of X, we have that for any $x \neq w$,

$$\varphi(w) = \limsup_{n \to \infty} \|x_n - w\| < \limsup_{n \to \infty} \|x_n - x\| = \varphi(x),$$

which implies that $\varphi(w) = \inf\{\varphi(x) : x \in C\}$. The conclusion follows from Theorem 3.3. $\square$

Every Hilbert space possesses the Opial property. Additionally, the sequence spaces l^p for $1 \leq p < \infty$ also exhibit the Opial property. However, it is important to note that many significant uniformly convex Banach spaces, such as L^p for $1 < p \neq 2$, do not possess the Opial property. As a result, another variant of the Demiclosedness Principle is required. In the version presented below, we will assume that X is both uniformly convex and uniformly smooth. Notably, L^p for $p > 1$, including the Hilbert space L^2, serve as prime examples of such spaces.

Theorem 3.5 (The Demiclosedness Principle 2) *Let X be a uniformly convex and uniformly smooth Banach space X. Let C be a nonempty, bounded, closed, and convex subset of X, and let $\mathcal{T} \in \mathcal{APNS}(C)$. Assume that there exists $w \in X$ and a regular sequence $\{x_n\}$ of elements of C such that $x_n \rightharpoonup w$. If $\{x_n\}$ is an approximate fixed point sequence for an operator $T_s \in \mathcal{T}$, then $w \in F(T_{ks})$ for any natural k.*

3.4 Asymptotic Pointwise Nonexpansive Sequences

Proof From the subdifferential inequality (3.1) in Proposition 3.1, we derive the following inequality which holds for any $x \in C$,

$$\frac{1}{2}\|x_n - w\|^2 + \langle w - x, J(x_n - w)\rangle \leq \frac{1}{2}\|x_n - x\|^2. \tag{3.18}$$

Since $x_n \rightharpoonup w$ as n approaches infinity, by taking the limit superior of both sides of (3.18), we find that $\varphi(w)^2 \leq \varphi(x)^2$, which implies that $\varphi(w) = \inf\{\varphi(x) : x \in C\}$, where φ is defined as in Theorem 3.4. The conclusion follows from Theorem 3.3. $\square$

Building on our discussion in Remark 3.3, and as a corollary to Theorems 3.4 and 3.5, we can present a classical version of The Demiclosedness Principle for nonexpansive mappings. This not only shows that our findings are genuine generalisations of established results in fixed point theory for nonexpansive mappings, but it will also be directly applicable in some common fixed point construction procedures discussed in the following chapters.

Theorem 3.6 (The Demiclosedness Principle 3) *Let X be a uniformly convex Banach space. Assume that either X has the Opial property or that X is uniformly smooth. Let $T : C \to C$ be a nonexpansive mapping, where C is a nonempty, bounded, closed, and convex subset of X. Assume that an approximate fixed point sequence $\{x_n\}$ converges in the weak topology to an element $w \in C$. Then, w is a fixed point of mapping T.*

3.4 Asymptotic Pointwise Nonexpansive Sequences

As we have already observed, and as will become increasingly evident in the following chapters, studying the construction of common fixed points for semigroups of operators is a challenging task. In simple terms, there are too many parameters to manage. For instance, if $J = [0, +\infty)$, then we need to serialise a continuous family of operators, which requires us to find a sequence $\{t_k\}$ of elements from J that ensures the constructed sequence $\{x_k\}$ of elements of C converges, in a suitable topology, to a common fixed point. Even when J is a discrete set, we may still need to identify a subsequence that ensures the convergence of the process. The asymptotic pointwise nonexpansive behaviour of the operators in these semigroups introduces an additional layer of difficulty. However, this challenge is largely counterbalanced by the increased flexibility in selecting a suitable set C, as discussed in Chap. 1.

To enhance simplicity and promote reusability across different variants of the construction processes, we will now introduce the concept of asymptotic pointwise nonexpansive sequences. This will be followed by some technically intricate yet significant results that will be employed in the subsequent chapters.

Definition 3.6 Let C be a subset of a Banach space X. We say that a sequence $\{T_n\}$ of mappings that act within C is an asymptotic pointwise nonexpansive sequence if for every $x \in C$ and every $n \in \mathbb{N}$ there exists an $A_n(x) > 0$ such that

$$\|T_n(x) - T_n(y)\| \leq A_n(x)\|x - y\| \tag{3.19}$$

holds for all $y \in C$, and such that for every $x \in C$,

$$\lim_{n \to \infty} A_n(x) = 1. \tag{3.20}$$

Define $B_n(x) = A_n(x) - 1$. In view of (3.20), we have

$$\lim_{n \to \infty} B_n(x) = 0.$$

By $\mathcal{A}(C)$ we will denote the class of all asymptotic pointwise nonexpansive sequences of mappings $\{T_n\}$ acting within C. Let $\{T_n\} \in \mathcal{A}(C)$. We say that a sequence $\{x_n\}$ of elements of C is generated by $\{T_n\}$ if $x_1 \in C$ and $x_{n+1} = T_n(x_n)$ for every $n \in \mathbb{N}$.

We start with the following technical results.

Lemma 3.5 *Let C be a subset of a Banach space X. Let $\{T_k\} \in \mathcal{A}(C)$ and let $\{x_k\}$ be a sequence generated by $\{T_k\}$. Assume that $w \in \bigcap_{k=1}^{\infty} F(T_k)$ is such that*

$$\sum_{i=1}^{\infty} B_i(w) < \infty. \tag{3.21}$$

There exists then an $r \in \mathbb{R}$ such that $\lim_{k \to \infty} \|x_k - w\| = r$.

Proof Take any $k, n \in \mathbb{N}$ and observe that

$$\|x_{k+1} - w\| = \|T_k(x_k) - T_k(w)\| \leq A_k(w)\|x_k - w\|.$$

This implies the next inequality:

$$\|x_{k+n} - w\| \leq \prod_{i=k}^{k+n-1} A_i(w)\|x_k - w\|. \tag{3.22}$$

By passing with $n \to \infty$ in (3.22), we find that

$$\limsup_{m \to \infty} \|x_m - w\| \leq \prod_{i=k}^{\infty} A_i(w)\|x_k - w\|. \tag{3.23}$$

It follows from (3.21) that $\prod_{i=k}^{\infty} A_i(w) \to 1$ as $k \to \infty$. Thus, by passing with k to infinity in (3.23), we arrive at the desired inequality,

3.4 Asymptotic Pointwise Nonexpansive Sequences

$$\limsup_{m\to\infty} \|x_m - w\| \leq \liminf_{k\to\infty} \|x_k - w\|.$$

□

Let us define a specific class of asymptotic pointwise nonexpansive sequences that meet certain beneficial control properties.

Definition 3.7 Define $\mathcal{A}_c(C)$ as a class of all $\{T_n\} \in \mathcal{A}(C)$ such that for every $x \in C$ there exists a positive constant ε_x, for which

$$\sum_{n=1}^{\infty} B_n(x_n) < \infty \qquad (3.24)$$

holds for every sequence $\{x_n\}$ such that $x_n \in C \cap B(x, \varepsilon_x) = \{u \in C : \|u - x\| \leq \varepsilon\}$.

Remark 3.4 Observe that a sequence $\{T_n\}$ belongs to $\mathcal{A}_c(C)$ if for every $x \in C$ there exists a positive constant ε_x and a sequence of finite positive constants $\{M_n(x)\}$ such that

$$\sum_{n=1}^{\infty} M_n(x) < \infty$$

and

$$B_n(y) \leq M_n(x)$$

hold for every $y \in C \cap B(x, \varepsilon_x)$. In certain cases, these conditions provide the simplest method to demonstrate that the sequence of mappings $\{T_n\}$ belongs to the class $\mathcal{A}_c(C)$.

Remark 3.5 Let $\{T_k\} \in \mathcal{A}_c(C)$ and $x \in C$. It immediately follows from Definition 3.7 that $\sum_{i=1}^{\infty} B_i(x) < \infty$. In particular, $\sum_{i=1}^{\infty} B_i(w) < \infty$, when $w \in \bigcap_{k=1}^{\infty} F(T_k)$.

Remark 3.6 Let $\{T_n\} \in \mathcal{A}(C)$. Observe that if the functions $A_n(x)$ for all mappings T_n do not depend on x, that is, they are all constants denoted by A_n, then the condition (3.19) of Definition 3.6 is replaced by

$$\|T_n(x) - T_n(y)\| \leq A_n \|x - y\|, \qquad (3.25)$$

where $1 \leq A_n \to 1$, and the condition (3.24) is replaced by a simpler inequality $\sum_{n=1}^{\infty} B_n < \infty$.

The next result is an immediate consequence of Lemma 3.5 and Remark 3.5.

Lemma 3.6 Let C be a subset of a Banach space X. Let $\{T_k\} \in \mathcal{A}_c(C)$ and let $\{x_k\}$ be a sequence generated by $\{T_k\}$. Assume that $w \in \bigcap_{k=1}^{\infty} F(T_k)$. There exists then an $r \in \mathbb{R}$ such that $\lim_{k\to\infty} \|x_k - w\| = r$.

Lemma 3.6 plays a crucial role in establishing our next technical result.

Lemma 3.7 *Let C be a bounded, closed, and convex subset of a uniformly convex Banach space X. Let $\{T_k\} \in \mathcal{A}_c(C)$ and let $\{x_k\}$ be a sequence generated by $\{T_k\}$. Assume that $w_1, w_2 \in \bigcap_{k=1}^{\infty} F(T_k)$. Then, there exists a real number $t_0 \in [0, 1]$ such that the limit*

$$r_t = \lim_{k \to \infty} \|tx_k + (1-t)w_1 - w_2\| \tag{3.26}$$

exists and is finite for any number $t \in [0, t_0]$.

Proof Denote

$$S_{k,0}(x) = x,$$

$$S_{k,m} = T_{k+m-1} \circ T_{k+m-2} \circ \ldots \circ T_k, \ for \ m \in \mathbb{N},$$

$$f_k(t) = \|tx_k + (1-t)w_1 - w_2\|.$$

Since C is convex and bounded, it follows that $\sup\{f_k(t) : t \in [0, 1], \ k \in \mathbb{N}\} < \infty$. We will also use the following useful notation:

$$g_{k,m}(t) = \|S_{k,m}(tx_k + (1-t)w_1) - (tx_{k+m} + (1-t)w_1)\|.$$

It is easy to calculate that $S_{k,m}(x_k) = x_{k+m}$. Since

$$\|T_k(u) - T_k(v)\| \le A_k(u)\|u - v\|,$$

it follows that

$$\|S_{k,m}(u) - S_{k,m}(v)\| \le \prod_{j=k}^{k+m-1} A_j\big(S_{k,j-k}(u)\big)\|u - v\|, \tag{3.27}$$

holds for every $u, v \in C$, which means that every $S_{k,m}$ is a pointwise Lipschitzian mapping in the sense of Definition 1.4. Denote $u_t^k = tx_k + (1-t)w_1$ and

$$h_{k,m}(u) = \prod_{j=k}^{k+m-1} A_j\big(S_{k,j-k}(u)\big).$$

Observe that $S_{k,m}(w_1) = w_1$. When this is considered alongside inequality (3.27) and the fact that $a_j \ge 1$, it leads to the inequality

$$\|S_{k,m}(u) - w_1\| = \|S_{k,m}(u) - S_{k,m}(w_1)\| \le \prod_{j=k}^{\infty} A_j(w_1)\|u - w_1\|. \tag{3.28}$$

Note also that

$$\|u_t^k - w_1\| = \|tx_k + (1-t)w_1 - w_1\| = t\|x_k - w_1\| \le t \operatorname{diam}(C). \tag{3.29}$$

3.4 Asymptotic Pointwise Nonexpansive Sequences

From (3.28) and (3.29) it follows that

$$\|S_{k,m}(u_t^k) - w_1\| \leq t \prod_{j=k}^{\infty} A_j(w_1) \, \mathrm{diam}(C). \tag{3.30}$$

Take ε_{w_1} as defined for the sequence $\{T_k\}$ in Definition 3.7. In view of (3.24), $\prod_{j=k}^{\infty} A_j(w_1) \to 1$ as $k \to \infty$. Thus, using (3.30), we can choose $k_0 \in \mathbb{N}$ and $t_0 > 0$ such that

$$\|S_{k,m}(u_t^k) - w_1\| \leq \varepsilon_{w_1} \tag{3.31}$$

for every $m \in \mathbb{N}$, every natural $k \geq k_0$ and every $t \in (0, t_0)$. It follows from (3.31) that each $S_{k,j-k}(u_t^k)$ belongs to the set $C \cap B(w_1, \varepsilon_{w_1})$. Therefore, utilising (3.24), we can conclude that

$$\sum_{j=k}^{\infty} B_j\left(S_{k,j-k}(u_t^k)\right) < \infty,$$

which implies that

$$h_k(u_t^k) := \prod_{j=k}^{\infty} A_j\left(S_{k,j-k}(u_t^k)\right) \to 1$$

as $k \to \infty$. Note that $0 < h_{k,m}(u_t^k) < \infty$ and that $h_{k,m}(u_t^k) \to h_k(u_t^k)$ as $m \to \infty$.

It follows from (3.27) that each $S_{k,m}$ is a pointwise Lipschitzian mapping with $\alpha(x) = h_{k,m}(x)$. Given that $S_{k,m}(x_k) = x_{k+m}$, $w_1 \in F(S_{k,m})$ and $u_t^k = tx_k + (1-t)w_1$, we can invoke Lemma 2.1 for $S_{k,m}$. This lemma enables us to compute $g_{k,m}(t)$ by

$$g_{k,m}(t) = \|S_{k,m}(tx_k + (1-t)w_1) - (tS_{k,m}(x_k) + (1-t)S_{k,m}(w_1))\| \tag{3.32}$$

$$\leq h_{k,m}(u_t^k)\gamma_2^{-1}\left(\|x_k - w_1\| - \frac{\|x_{k+m} - w_1\|}{h_{k,m}(u_t^k)}\right).$$

Since γ_2^{-1} is an increasing function and $h_{k,m}(u_t^k) \geq 1$, it follows easily from (3.32) that

$$g_{k,m}(t) \leq h_{k,m}(u_t^k)\gamma_2^{-1}\Bigg(\|x_k - w_1\| - \|x_{k+m} - w_1\|$$

$$+ \left(1 - \frac{1}{h_{k,m}(u_t^k)}\right)\mathrm{diam}(C)\Bigg). \tag{3.33}$$

Denote $P_{k,m}(t) = x_{k+m} + (1-t)w_1 - w_2 - S_{k,m}(tx_k + (1-t)w_1)$. It is straightforward to check that $\|P_{k,m}(t)\| = g_{k,m}(t)$ and that

$$x_{k+m} + (1-t)w_1 - w_2 = P_{k,m}(t) + S_{k,m}(tx_k + (1-t)w_1) - w_2.$$

Therefore, we have the inequality

$$
\begin{aligned}
f_{k+m}(t) &= \|tx_{k+m} + (1-t)w_1 - w_2\| \quad &(3.34)\\
&= \|P_{k,m}(t) + S_{k,m}(tx_k + (1-t)w_1) - w_2\|\\
&\leq g_{k,m}(t) + \|S_{k,m}(tx_k + (1-t)w_1) - w_2\|\\
&\leq g_{k,m}(t) + \prod_{j=k}^{k+m-1} A_{n_j}(w_2)\|tx_k + (1-t)w_1 - w_2\|\\
&\leq g_{k,m}(t) + G_k f_k(t),
\end{aligned}
$$

where $0 < G_k := \prod_{j=k}^{\infty} A_{n_j}(w_2) < \infty$. Note that $G_k \to 1$. Let $r = \lim_{n\to\infty}\|x_n - w_1\|$, which exists by Lemma 3.6. After incorporating (3.33) into (3.34) we get

$$f_{k+m}(t) \leq h_{k,m}(u_t^k)\gamma_2^{-1}\left(\|x_k - w_1\| - \|x_{k+m} - w_1\| + \left(1 - \frac{1}{h_{k,m}(u_t^k)}\right)\text{diam}(C)\right) + G_k f_k(t).$$

Passing with $m \to \infty$ in both sides of this inequality and using continuity of γ_2^{-1} we arrive at the formula

$$\limsup_{n\to\infty} f_n(t) \leq h_k(u_t^k)\gamma_2^{-1}\left(\|x_k - w_1\| - r + \left(1 - \frac{1}{h_k(u_t^k)}\right)\text{diam}(C)\right) + G_k f_k(t).$$

Let $k \to \infty$. Considering that $\|x_k - w_1\| \to r$, $h_k(u_t^k) \to 1$, $G_k \to 1$, $\gamma_2^{-1}(0) = 0$ and γ_2^{-1} is continuous, we can derive the inequality

$$\limsup_{n\to\infty} f_n(t) \leq \liminf_{k\to\infty} f_k(t) < \infty.$$

We conclude that $\lim_{k\to\infty} f_k(t) = \lim_{k\to\infty}\|tx_k + (1-t)w_1 - w_2\|$ exists and is finite, as claimed. $\square$

The following lemma introduces the key technique for proving the weak convergence of the iterative processes in uniformly smooth Banach spaces.

Lemma 3.8 *Let C be a bounded, closed, and convex subset of a uniformly convex and uniformly smooth Banach space X. Let $\{T_k\} \in \mathcal{A}_c(C)$ and let $\{x_k\}$ be a sequence generated by $\{T_k\}$. Assume that*

$$w_1, w_2 \in \bigcap_{k=1}^{\infty} F(T_k).$$

Then,

$$\langle y - z, J(w_1 - w_2)\rangle = 0 \quad (3.35)$$

3.4 Asymptotic Pointwise Nonexpansive Sequences

holds for any two weak cluster points y, z of the sequence $\{x_k\}$.

Proof It follows from the definition of the weak convergence that, to establish equality (3.35), it suffices to demonstrate that the limit

$$\lim_{n\to\infty} \langle x_n, J(w_1 - w_2) \rangle$$

exists for any two fixed points $w_1, w_2 \in \bigcap_{k=1}^{\infty} F(T_k)$. To prove its existence, fix temporarily any $t \in (0, t_0]$, where t_0 is selected so (3.26) from Lemma 3.7 holds. Apply inequality (3.1) from Proposition 3.1 with $x = w_1 - w_2$ and $h = t(x_n - w_1)$ to obtain

$$\frac{1}{2} \|w_1 - w_2\|^2 + \langle t(x_n - w_1), J(w_1 - w_2) \rangle$$
$$\leq \frac{1}{2} \|tx_n + (1-t)w_1 - w_2\|^2 \quad (3.36)$$
$$\leq \frac{1}{2} \|w_1 - w_2\|^2 + \langle t(x_n - w_1), J(w_1 - w_2) \rangle + \|t(x_n - w_1)\|^2.$$

By passing with n to infinity in (3.36), while using Lemma 3.7 and the boundedness of C, we derive the inequalities

$$\frac{1}{2} |w_1 - w_2|^2 + t \limsup_{n\to\infty} \langle (x_n - w_1), J(w_1 - w_2) \rangle$$
$$\leq \frac{1}{2} \lim_{n\to\infty} \|tx_n + (1-t)w_1 - w_2\|^2 \quad (3.37)$$
$$\leq \frac{1}{2} \|w_1 - w_2\|^2 + t \liminf_{n\to\infty} \langle x_n - w_1, J(w_1 - w_2) \rangle + t^2 \text{diam}(C)^2.$$

From (3.37) we conclude that

$$\limsup_{n\to\infty} \langle (x_n - w_1), J(w_1 - w_2) \rangle \leq \liminf_{n\to\infty} \langle x_n - w_1, J(w_1 - w_2) \rangle + t \, \text{diam}(C)^2,$$

and, by the arbitrariness of $0 < t \leq t_0$, that

$$\limsup_{n\to\infty} \langle (x_n - w_1), J(w_1 - w_2) \rangle \leq \liminf_{n\to\infty} \langle x_n - w_1, J(w_1 - w_2) \rangle,$$

which ensures that the limit $\lim_{n\to\infty} \langle x_n, J(w_1 - w_2) \rangle$ exists. The proof is complete. $\square$

3.5 Notes

Unless specifically stated, the results of this chapter related to the Opial property follow the ideas from the 2013 paper by Kozlowski and Sims [77]. The results related to the asymptotic pointwise nonexpansive sequences and uniformly smooth Banach spaces are based on the 2012 paper by Kozlowski [62]. Several ideas used in this chapter can be traced back to the 1994 article by Tan and Xu [105] who studied convergence of Mann and Ishikawa process for asymptotically nonexpansive mappings, as defined by Goebel and Kirk [40].

Chapter 4
Generalised Krasnosel'skii-Mann Iteration Processes

Abstract In the 1950s, Mann (Proc. Am. Math. Soc. 4:506–510, 1953) and Krasnosel'skii (Uspiehi Mat. Nauk 10:123–127, 1955) independently laid the groundwork for a highly effective successive iteration method used in constructing fixed points. For a nonexpansive mapping $T : C \to C$, this method is defined as follows: an initial point $x_1 \in C$ is chosen arbitrarily, and the subsequent points are given by $x_{k+1} = c_k T(x_k) + (1 - c_k)x_k$, where $\{c_k\}$ is a sequence of numbers from the interval $(0, 1)$. In this chapter, we adapt the Krasnosel'skii-Mann method to the setting of pointwise Lipschitzian semigroups.

4.1 Preliminaries

It was recognised early on that the method of Picard's iterates and its generalisations—successful in constructing fixed points of contractions (as proven by Theorem 2.1 and noted in Remark 2.3)—do not generally yield a converging process in the nonexpansive case. Take, for example, the operator T defined as a rotation of the unit disc in $\mathbb{R}^2$ by an angle $0 < \alpha < \pi$ and begin iterating $T^k(x_1)$, where x_1 is any point of the unit disc different from the origin. This gap led Mann [82] and Krasnosel'skii [78] to independently establish, in the 1950s, the foundations of a very successful successive iteration method. For a nonexpansive mapping $T : C \to C$, this method is defined as follows: an initial point $x_1 \in C$ is chosen arbitrarily, and the subsequent points are given by $x_{k+1} = c_k T(x_k) + (1 - c_k)x_k$, where $\{c_k\}$ is a sequence of numbers from the interval $(0, 1)$. Consider two key features of this process:

1. The process guarantees that the generated sequence $\{x_k\}$ remains within the set C, owing to the convexity of C.
2. The step size $\|x_{k+1} - x_k\|$ is a fraction of $\|T(x_k) - x_k\|$. This property is essential for establishing the convergence of the process.

To extend the Krasnosel'skii-Mann method to the framework of pointwise Lipschitzian semigroups, while maintaining the key features outlined above, we begin by precisely defining the generalised Krasnosel'skii-Mann iteration process applicable to semigroups of nonlinear mappings.

Definition 4.1 Let $\mathcal{T} \in \mathcal{APNS}(C)$. Let $\{t_k\}$ be a sequence of elements in J and $\{c_k\}$ be a sequence of numbers from $(0, 1)$. The generalised Krasnosel'skii-Mann iteration process $gKM(\mathcal{T}, \{c_k\}, \{t_k\})$ generated by the semigroup $\mathcal{T}$, the sequences $\{c_k\}$ and $\{t_k\}$, is defined by the iterative formula

$$x_{k+1} = c_k T_{t_k}(x_k) + (1 - c_k)x_k, \tag{4.1}$$

where $x_1 \in C$ is chosen arbitrarily, and

(i) $\{c_k\}$ is bounded away from 0 and 1,
(ii) $\sum_{k=1}^{\infty} b_{t_k}(x) < \infty$ for every $x \in C$.

Recall that $b_t(x) = a_t(x) - 1$, where $a_t(x) \geq 1$ is chosen so that the inequality

$$\|T_t(x) - T_t(y)\| \leq a_t(x)\|x - y\|$$

holds for any $y \in C$.

Remark 4.1 It is important to note that under assumption (ii) of Definition 4.1, for every $x \in C$ we have $\lim_{n \to \infty} b_{t_n}(x) = 0$. This implies that $\lim_{n \to \infty} a_{t_n}(x) = 1$. In case when $t_n \to \infty$, the latter equality follows directly from the fact that $\mathcal{T}$ is asymptotic pointwise nonexpansive. However, we do not need to assume that $t_n \to \infty$ provided the sequence $\{t_n\}$ defining the Krasnosel'skii-Mann process satisfies (ii). This flexibility allows for the direct application of our framework, designed for more complex asymptotic pointwise nonexpansive semigroups, to a much simpler nonexpansive case, where $a_t(x) = 1$ for every $x \in C$ and all $t \in J$. See also discussion in Remark 4.4.

Remark 4.2 Since $\lim_{n \to \infty} a_{t_n}(x) = 1$, it is evident that the sequence of operators $\{T_{t_k}\}$, defined by the process $gKM(\mathcal{T}, \{c_k\}, \{t_k\})$, constitutes an asymptotic pointwise nonexpansive sequence as per Definition 3.6, that is, $\{T_k\} \in \mathcal{A}(C)$.

Definition 4.2 We say that a generalised Krasnosel'skii-Mann iteration process $gKM(\mathcal{T}, \{c_k\}, \{t_k\})$ is well-defined if the sequence $\{x_k\}$, generated by it, is regular with respect to $\mathcal{T}$, and for any $s \in J$,

$$\lim_{k \to \infty} a_{s+t_k}(x_k) = 1. \tag{4.2}$$

Remark 4.3 Observe that by the definition of asymptotic pointwise nonexpansiveness, $\lim_{t \to \infty} a_t(x) = 1$ for every $x \in C$. Hence we can always select a sequence $\{t_k\}$ such that (4.2) holds. In other words, by a suitable choice of $\{t_k\}$ we can always make $gKM(\mathcal{T}, \{c_k\}, \{t_k\})$ well-defined. Furthermore, for many semigroups, such as asymptotic nonexpansive or nonexpansive semigroups, every generalised Krasnosel'skii-Mann iteration process is automatically well-defined.

4.1 Preliminaries

Remark 4.4 Following our previous observations in Remarks 3.3, 4.1, and Theorem 3.6, let us interpret the above definitions for the case of a semigroup $\mathcal{T} = \{T^k\}$, where T is a nonexpansive operator. In this scenario, $t_k = 1$ for every natural number k, condition (ii) of Definition 4.1 is automatically satisfied, and the process is well-defined. Furthermore, since $T_{t_k} = T$ for any $k \in \mathbb{N}$, the iterative formula (4.1) takes the form

$$x_{k+1} = c_k T(x_k) + (1 - c_k)x_k,$$

which represents the classical and standard form of the Krasnosel'skii-Mann iteration process for constructing a fixed point of a nonexpansive operator T.

We will establish a series of lemmas that are essential for proving the convergence theorems of the generalised Krasnosel'skii-Mann process.

Lemma 4.1 *Let C be a bounded, closed, and convex subset of a Banach space X. Let $\mathcal{T} \in \mathcal{APNS}(C)$, $w \in F(\mathcal{T})$, and let $\{x_k\}$ be a sequence generated by a generalised Krasnosel'skii-Mann process $gKM(\mathcal{T}, \{c_k\}, \{t_k\})$. Then, there exists an $r \in \mathbb{R}$ such that $\lim_{k \to \infty} \|x_k - w\| = r$.*

Proof Observe that the sequence $\{T_{t_k}\}$ is an asymptotic pointwise nonexpansive sequence in the sense of Definition 3.6 (recall Remark 4.2). Let $w \in F(\mathcal{T})$. Since $\sum_{i=1}^{+\infty} b_{t_i}(w) < \infty$, it follows from Lemma 3.5 that there exists an $r \in \mathbb{R}$ such that $\lim_{k \to \infty} \|x_k - w\| = r$, as claimed. □

Lemma 4.2 *Let C be a bounded, closed, and convex subset of a uniformly convex Banach space X. Let $\mathcal{T} \in \mathcal{APNS}(C)$. Let $\{x_k\}$ be a sequence generated by a well-defined generalised Krasnosel'skii-Mann process $gKM(\mathcal{F}, \{c_k\}, \{t_k\})$. Then,*

$$\lim_{k \to \infty} \|T_{t_k}(x_k) - x_k\| = 0 \tag{4.3}$$

and

$$\lim_{k \to \infty} \|x_{k+1} - x_k\| = 0. \tag{4.4}$$

Proof From Theorem 2.3 it follows that $F(\mathcal{T}) \neq \emptyset$. Let us fix $w \in F(\mathcal{T})$. By Lemma 4.1 there exists an $r \in \mathbb{R}$ such that $\lim_{k \to \infty} \|x_k - w\| = r$. Since $w \in F(\mathcal{T})$, and the process is well-defined, it follows that

$$\limsup_{k \to \infty} \|T_{t_k}(x_k) - w\| = \limsup_{k \to \infty} \|T_{t_k}(x_k) - T_{t_k}(w)\| \leq \limsup_{k \to \infty} a_{t_k}(x_k)\|x_k - w\| = r.$$

Observe that

$$\lim_{k \to \infty} \|c_k(T_{t_k}(x_k) - w) + (1 - c_k)(x_k - w)\| = \lim_{k \to \infty} \|x_{k+1} - w\| = r.$$

By applying Lemma 3.3 to $u_k = x_k - w$ and $v_k = T_{t_k}(x_k) - w$, we conclude that $\lim_{k \to \infty} \|T_{t_k}(x_k) - x_k\| = 0$, which, due to the definition of the sequence $\{x_k\}$, is equivalent to $\lim_{k \to \infty} \|x_{k+1} - x_k\| = 0$. □

The next lemma is an important technical result that shows that, under suitable assumptions, the sequence $\{x_k\}$, generated by the generalised Krasnosel'skii-Mann iteration process, is an approximate fixed point sequence for the underlying semigroup.

Lemma 4.3 *Let C be a bounded, closed, and convex subset of a uniformly convex Banach space X. Let $\mathcal{T} \in \mathcal{APNS}(C)$. Assume that the generalised Krasnosel'skii-Mann process $gKM(\mathcal{T}, \{c_k\}, \{t_k\})$ is well defined. Let $A \subset J$ be such that to every $s \in A$ there exists a strictly increasing sequence of natural numbers $\{j_k\}$ satisfying the following conditions:*

$$\lim_{k \to \infty} \|x_k - x_{j_k}\| = 0, \tag{4.5}$$

$$e_k := |t_{j_{k+1}} - t_{j_k} - s| \in J, \text{ for all } k \in \mathbb{N}, \tag{4.6}$$

and

$$\lim_{k \to \infty} \|T_{e_k}(x_{j_k}) - x_{j_k}\| = 0. \tag{4.7}$$

Then, $\{x_k\}$ is an approximate fixed point sequence for all mappings $\{T_{ms}\}$ where $s \in A$ and $m \in \mathbb{N}$, that is

$$\lim_{k \to \infty} \|T_{ms}(x_k) - x_k\| = 0 \tag{4.8}$$

for every $s \in A$ and $m \in \mathbb{N}$. If, in addition, A is a generating set for J then

$$\lim_{k \to \infty} \|T_t(x_k) - x_k\| = 0$$

for any $t \in J$.

Proof According to Lemma 3.4, it suffices to establish (4.8) for $m = 1$. To achieve this, fix any $s \in A$. Note that

$$\|x_{j_k} - x_{j_{k+1}}\| \to 0 \text{ as } k \to \infty, \tag{4.9}$$

which follows from

$$\|x_{j_k} - x_{j_{k+1}}\| \leq \|x_{j_k} - x_k\| + \|x_k - x_{k+1}\| + \|x_{k+1} - x_{j_{k+1}}\| \to 0. \tag{4.10}$$

The convergence to zero in (4.10) is a result of the assumption (4.5) combined with the equality (4.4) demonstrated in Lemma 4.2. Observe that the expression

$$\|x_{j_k} - T_s(x_{j_k})\| \leq \|x_{j_k} - x_{j_{k+1}}\| + \|x_{j_{k+1}} - T_{t_{j_{k+1}}}(x_{j_{k+1}})\| + \|T_{t_{j_{k+1}}}(x_{j_{k+1}}) - T_{t_{j_{k+1}}}(x_{j_k})\|$$
$$+ \|T_{t_{j_{k+1}}}(x_{j_k}) - T_{s+t_{j_k}}(x_{j_k})\| + \|T_{s+t_{j_k}}(x_{j_k}) - T_s(x_{j_k})\|$$
$$\leq \|x_{j_k} - x_{j_{k+1}}\| + \|x_{j_{k+1}} - T_{t_{j_{k+1}}}(x_{j_{k+1}})\| + a_{t_{j_{k+1}}}(x_{j_{k+1}})\|x_{j_{k+1}} - x_{j_k}\|$$
$$+ a_{s+t_{j_k}}(x_{j_k})\|T_{e_k}(x_{j_k}) - x_{j_k}\| + \sup_{n \in \mathbb{N}} a_s(x_{j_n})\|T_{t_{j_k}}(x_{j_k}) - x_{j_k}\|$$

tends to the zero as $k \to \infty$. Indeed, the first term tends to zero due to (4.9), while the second term converges to zero by (4.3) of Lemma 4.2. The third term tends to zero based on (4.9) and the fact that the process is well-defined. Similarly, the fourth term tends to zero by assumption (4.7) and, again, due to the well defined nature of the process. Finally, the last term converges to zero by (4.3) from Lemma 4.2 and the regularity of the sequence $\{x_k\}$. From the above, we conclude that

$$\|x_{j_k} - T_s(x_{j_k})\| \to 0 \text{ as } k \to \infty. \tag{4.11}$$

Similarly, using (4.11) and the estimation

$$\begin{aligned}\|x_k - T_s(x_k)\| &\leq \|x_k - x_{j_k}\| + \|x_{j_k} - T_{t_{j_k}}(x_{j_k})\| + \|T_{t_{j_k}}(x_{j_k}) - T_{s+t_{j_k}}(x_{j_k})\| \\ &\quad + \|T_{s+t_{j_k}}(x_{j_k}) - T_s(x_{j_k})\| + \|T_s(x_{j_k}) - T_s(x_k)\| \\ &\leq \|x_k - x_{j_k}\| + \|x_{j_k} - T_{t_{j_k}}(x_{j_k})\| + a_{t_{j_k}}(x_{j_k})\|x_{j_k} - T_s(x_{j_k})\| \\ &\quad + a_s(x_{j_k})\|T_{j_k}(x_{j_k}) - x_{j_k}\| + a_s(x_k)\|x_{j_k} - x_k\|,\end{aligned}$$

we conclude that $\|x_k - T_s(x_k)\| \to 0$ as $k \to \infty$. This means that $\{x_k\}$ is an approximate fixed point sequence for the mapping $\{T_s\}$. If A is a generating set for J, then by Theorem 3.1, $\{x_k\}$ is an approximate fixed point sequence for any $T_s \in \mathcal{T}$. This concludes the proof of the lemma. $\square$

Remark 4.5 Observe that when $J = [0, +\infty)$, the assumption given in (4.6) of Lemma 4.3 is automatically satisfied. Therefore, in these instances, verification of this assumption is unnecessary.

4.2 Weak Convergence of Generalised Krasnosel'skii-Mann Iteration Processes in Uniformly Convex Banach Spaces with the Opial Property

In this section, we assume that the uniformly convex Banach space X has the Opial property. We will prove a generic version of the weak convergence theorem for the sequences $\{x_k\}$ that are generated by the Krasnosel'skii-Mann iteration process and also serve as approximate fixed point sequences. This will be followed by the discussion of several applications of the generic convergence theorem to some more specific situations.

Theorem 4.1 *Let X be a uniformly convex Banach space X with the Opial property. Let C be a bounded, closed, and convex subset of a X, and let $\mathcal{T} \in \mathcal{APNS}(C)$. Let A be a generating set for J. Assume that $gKM(\mathcal{T}, \{c_k\}, \{t_k\})$ is a well-defined Krasnosel'skii-Mann iteration process. If the sequence $\{x_k\}$ generated by $gKM(\mathcal{T}, \{c_k\}, \{t_k\})$ is an approximate fixed point sequence for every T_s, where $s \in A$, then there exists $w \in F(\mathcal{T})$ such that $x_k \rightharpoonup w$.*

Proof By Theorem 3.1, $\{x_k\}$ is an approximate fixed point sequence for any $s \in J$. Consider $y, z \in C$, which are two weak cluster points of the sequence $\{x_k\}$. Then, there exist two subsequences $\{y_k\}$ and $\{z_k\}$ of $\{x_k\}$ such that $y_k \rightharpoonup y$ and $z_k \rightharpoonup z$. It follows from the Demiclosedness Principle 1 (Theorem 3.4) that $T_s(y) = y$ and $T_s(z) = z$ for all $s \in J$, which means that $y, z \in F(\mathcal{T})$. By Lemma 4.1 the following limits exist

$$r_1 = \lim_{k \to \infty} \|x_k - y\|, \quad r_2 = \lim_{k \to \infty} \|x_k - z\|.$$

We claim that $y = z$. Assume, for the sake of contradiction, that $y \neq z$. By the Opial property,

$$r_1 = \liminf_{k \to \infty} \|y_k - y\| < \liminf_{k \to \infty} \|y_k - z\| = r_2 \quad (4.12)$$
$$= \liminf_{k \to \infty} \|z_k - z\| < \liminf_{k \to \infty} \|z_k - y\| = r_1.$$

The contradiction implies $y = z$ which means that the sequence $\{x_k\}$ has at most one weak cluster point. Since C is weakly sequentially compact, we deduce that the sequence $\{x_k\}$ has exactly one weak cluster point $w \in C$, which means that $x_k \rightharpoonup w$. By applying the Demiclosedness Principle again, we conclude that $T_t(w) = w$ for any $t \in J$, as claimed. □

Let us apply the above result to some more specific situations. Let us start with a discrete case. First, we need to recall the following notions.

Definition 4.3 A strictly increasing sequence of natural numbers $\{n_i\}$ is called quasi-periodic if the sequence $\{n_{i+1} - n_i\}$ is bounded, or equivalently, if there exists a number $p \in \mathbb{N}$ such that any block of p consecutive natural numbers must contain a term of the sequence $\{n_i\}$. The smallest of such numbers p will be called a quasi-period of $\{n_i\}$.

Theorem 4.2 *Let X be a uniformly convex Banach space X with the Opial property, and let C be a bounded, closed, and convex subset of X. Assume that the semigroup $\mathcal{T} \in \mathcal{APNS}(C)$ is parametrised by a subsemigroup J of the additive semigroup $\mathbb{N}_0$, which implies that J is generated by a finite set $A = \{\alpha_1, \alpha_2, \alpha_3 ..., \alpha_M\}$ where each α_i is a natural number (recall Remark 3.2). Assume that $gKM(\mathcal{T}, \{c_k\}, \{t_k\})$ is a well-defined Krasnosel'skii-Mann iteration process. Assume that for every natural number $m \leq M$, there exists a strictly increasing, quasi-periodic sequence of natural numbers $\{j_k(m)\}$, with a quasi-period p_m, such that the equality*

$$t_{j_{k+1}(m)} = \alpha_m + t_{j_k(m)}$$

holds for every $k \in \mathbb{N}$. Then, the sequence $\{x_k\}$, generated by $gKM(\mathcal{T}, \{c_k\}, \{t_k\})$, is an approximate fixed point sequence for every T_s, where $s \in A$. Furthermore, $\{x_k\}$ converges weakly to a common fixed point $w \in F(\mathcal{T})$.

4.2 Weak Convergence of Generalised Krasnosel'skii-Mann ...

Proof We will apply Lemma 4.3. Note that the assumptions (4.6) and (4.7) of Lemma 4.3 are trivially satisfied since $t_{j_{k+1}(m)} - t_{j_k(m)} - \alpha_m = 0$. To prove (4.5) of the same lemma, observe that by the quasi-periodicity of $\{j_k(m)\}$, to every positive integer k there exists $j_k(m)$ such that $|k - j_k(m)| \leq p_m$. Assume that $k - p_m \leq j_k(m) \leq k$ (the proof for the other case is identical). Fix $\varepsilon > 0$. Note that by Lemma 4.2, $\|x_{k+1} - x_k\| < \frac{\varepsilon}{p_m}$ for k sufficiently large. Hence for k sufficiently large,

$$\|x_k - x_{j_k}\| \leq \|x_k - x_{k-1}\| + \ldots + \|x_{j_k(m)+1} - x_{j_k(m)}\| \leq p_m \frac{\varepsilon}{p_m} = \varepsilon.$$

This proves that (4.5) is also satisfied. Therefore, by Lemma 4.3, $\{x_k\}$ is an approximate fixed point sequence for every T_s where $s \in A$. By Theorem 4.1 (see also Remark 3.2), $\{x_k\}$ converges weakly to a common fixed fixed point $w \in F(\mathcal{T})$. □

Remark 4.6 It is clear that we can always construct a sequence $\{t_k\}$ with the properties specified in the assumptions of Theorem 4.2. For example, one can take a concatenation of all sequences $\{k\alpha_i\}_{i=1}^M$, where k runs over the set all natural numbers. When implementing this algorithm concretely, the challenge will be to ensure that with the chosen sequence $\{t_k\}$ the Krasnosel'skii-Mann process $gKM(\mathcal{T}, \{c_k\}, \{t_k\})$ remains well-defined (see Definition 4.2). However, let us recall that for many semigroups, such as asymptotic nonexpansive or nonexpansive semigroups, every Krasnosel'skii-Mann iteration process is automatically well-defined.

Remark 4.7 Note that by setting $A = \{1\}$, Theorem 4.2 demonstrates the weak convergence of the Krasnosel'skii-Mann iteration processes for asymptotic pointwise nonexpansive mappings and, consequently for nonexpansive mappings. The case of nonexpansive mappings warrants a closer examination, as it demonstrates that Theorem 4.2 is a genuine generalisation of the classical Krasnosel'skii-Mann iteration method for finding a fixed point of a single nonexpansive operator. This is addressed in the following result.

Theorem 4.3 *Let X be a uniformly convex Banach space X with the Opial property, and let C be a bounded, closed, and convex subset of X. Suppose that $T : C \to C$ is nonexpansive and consider the Krasnosel'skii-Mann process defined by*

$$x_{k+1} = c_k T(x_k) + (1 - c_k) x_k$$

(recall Remark 4.4). Then, the sequence $\{x_k\}$ generated by this process is an approximate fixed point sequence for T. Furthermore, $\{x_k\}$ converges weakly to a fixed point of T.

Proof Observe that in this scenario, $J = \mathbb{N}_0$, $A = \{1\}$, $t_k = 1$. Therefore, all assumptions of Theorem 4.2 pertaining to $\{t_k\}$ are satisfied. □

The following useful result is a consequence of Lemma 4.3, Theorems 3.2, and 4.1. Its practical significance is underscored by the fact the set B can be made countable, simplifying computational implementations considerably.

Theorem 4.4 *Let X be a uniformly convex Banach space X with the Opial property. Let C be a bounded, closed, and convex subset of X. Let $\mathcal{T} \in \mathcal{APNS}(C)$ be a semigroup with $J = [0, +\infty)$. Assume that $\mathcal{T}$ is an equicontinuous semigroup. Let $\{x_k\}$ be the sequence generated by a well-defined generalised Krasnosel'skii-Mann iteration process $gKM(\mathcal{T}, \{c_k\}, \{t_k\})$. Assume that $B \subset A = [0, 1)$ is dense in A. If to every $s \in B$ there exists a strictly increasing sequence of natural numbers $\{j_k\}$ satisfying the conditions*

(a) $t_{j_{k+1}} - t_{j_k} \to s$ as $k \to \infty$,
(b) $\|x_k - x_{j_k}\| \to 0$ as $k \to \infty$,

then the sequence $\{x_k\}$ converges weakly to a common fixed point $w \in F(\mathcal{T})$.

Proof To apply Lemma 4.3, we note that the condition (4.5) in this lemma is satisfied due to our assumption (b), while (4.7) follows readily from (a) through the equicontinuity of $\mathcal{T}$. Thus, by Lemma 4.3, the sequence $\{x_k\}$ is an approximate fixed point sequence for all mappings $\{T_s\}$ where $s \in B$. Since B is a dense subset of A, it follows from Theorem 3.2 that sequence $\{x_k\}$ is an approximate fixed point sequence for mappings $\{T_s\}$ for all $s \in A$. Since $A = [0, 1)$ is a generating set for $J = [0, +\infty)$, our assertion follows directly from Theorem 4.1. □

Our next result demonstrates that, in certain cases, the task of finding a common fixed point for all mappings in $\mathcal{T}$ can be simplified to the problem of finding a common fixed point for only two specific mappings.

Theorem 4.5 *Let X be a uniformly convex Banach space X with the Opial property. Let C be a bounded, closed, and convex subset of X. Let $\mathcal{T} \in \mathcal{APNS}(C)$ be equicontinuous and let $J = [0, +\infty)$. Let $\{x_k\}$ be generated by a well defined Krasnosel'skii-Mann iteration process $gKM(\mathcal{T}, \{c_k\}, \{t_k\})$. Let $s_1, s_2 \in J$ be such that the ratio s_1/s_2 is irrational and that there exist two strictly increasing sequence of natural numbers $\{i_k\}$ and $\{j_k\}$ satisfying the conditions*

$$\lim_{k \to \infty} \|x_k - x_{i_k}\| = 0,$$

$$\lim_{k \to \infty} \|x_k - x_{j_k}\| = 0,$$

$$\lim_{k \to \infty} \|T_{e_k}(x_{j_k}) - x_{j_k}\| = 0,$$

$$\lim_{k \to \infty} \|T_{f_k}(x_{j_k}) - x_{j_k}\| = 0,$$

where $e_k = |t_{i_{k+1}} - t_{i_k} - s_1|$ and $f_k = |t_{j_{k+1}} - t_{j_k} - s_2|$, then the sequence $\{x_k\}$ converges weakly to a common fixed point $w \in F(\mathcal{T})$.

Proof It follows from Lemma 4.3 that $\{x_k\}$ is an approximate fixed point sequence for both T_{s_1} and T_{s_2}. By Theorem 4.1 there exists an element $w \in C$ such that $x_k \rightharpoonup w$, $T_{s_1}(w) = w$ and $T_{s_2}(w) = w$. Since the ratio s_1/s_2 is irrational, it follows from Proposition 3.2 that $w \in F(\mathcal{T})$. □

4.3 Weak Convergence of Generalised Krasnosel'skii-Mann Iteration Processes in Uniformly Convex and Uniformly Smooth Banach Spaces

In this section, we assume that the uniformly convex Banach space X also possesses the property of uniform smoothness. As noted in Chap. 3, many uniformly convex Banach spaces, such as L^p for $1 < p \neq 2$, lack the Opial property while enjoying the uniform smoothness. This observation underscores the importance of the weak convergence results discussed in this section.

Following the approach used for spaces with the Opial property, we will establish a generic version of the weak convergence theorem for the sequences $\{x_k\}$ that are generated by the Krasnosel'skii-Mann iteration process and are also approximate fixed point sequences. Afterward, we will explore various applications of this general convergence theorem in more specific contexts.

To fully leverage the advantages of uniform smoothness, we will narrow our focus in this section to a subclass of the generalised Krasnosel'skii-Mann processes, specifically those that are locally uniformly bounded, defined as follows.

Definition 4.4 A generalised Krasnosel'skii-Mann iteration process is said to be locally uniformly bounded, and denoted as $gKM_{UB}(\mathcal{T}, \{c_k\}, \{t_k\})$, if the assumption

(ii) $\sum_{n=1}^{\infty} b_{t_n}(x) < \infty$ for every $x \in C$.

of Definition 4.1 is replaced by the condition

(ii') For every $x \in C$ there exists a positive constant ε_x and a series of positive constants $\{M_n(x)\}$ such that

$$\sum_{n=1}^{\infty} M_n(x_n) < \infty,$$

and

$$b_{t_n}(y) \leq M_n(x)$$

holds for every $y \in C \cap B(x, \varepsilon_x)$.

Clearly, every locally uniformly bounded generalised Krasnosel'skii-Mann iteration process qualifies as a generalised Krasnosel'skii-Mann iteration process according to Definition 4.1. Therefore, all previous results concerning generalised Krasnosel'skii-Mann iteration processes also apply to their locally uniformly bounded counterparts. Condition (ii) has been introduced to ensure the sequence $\{t_k\}$ is selected such that the sum of all remainders, defined as $b_{t_n}(x) = a_{t_n}(x) - 1$, is pointwise finite. The new condition (ii') requires that this series be locally uniformly summable. While this condition may initially seem unusual, it is reasonable to expect some form of controlled behaviour for such series, particularly since we assume that $a_{t_n}(x) \to 1$ as $n \to \infty$. Additionally, it is worth noting that conditions (ii) and (ii')

are equivalent when all functions $a_t(\cdot)$ are constant, which occurs when the semigroup is asymptotically nonexpansive (and in particular—nonexpansive). More generally, this equivalence holds if each of the functions $a_t(\cdot)$ is a weak-continuous function on C.

For the sake of brevity, in the remainder of this section, we will use the notation $gKM_{UB}(\mathcal{T}, \{c_k\}, \{t_k\})$, which will always refer to a locally uniformly bounded generalised Krasnosel'skii-Mann iteration process, even when the phrase "locally uniformly bounded" is omitted.

Lemma 4.4 *Let C be a bounded, closed, and convex subset of a uniformly convex and uniformly smooth Banach space X. Let $\mathcal{T} \in \mathcal{APNS}(C)$ and let $gKM_{UB}(\mathcal{T}, \{c_k\}, \{t_k\})$ be a well defined generalised Krasnosel'skii-Mann iteration process. Then, for every $w_1, w_2 \in F(\mathcal{T})$ and for any two weak cluster points y, z of the sequence $\{x_k\}$ generated by $gKM_{UB}(\mathcal{T}, \{c_k\}, \{t_k\})$,*

$$\langle y - z, J(w_1 - w_2) \rangle = 0. \tag{4.13}$$

Proof Let $w_1, w_2 \in F(\mathcal{T})$. Define

$$H_k(u) = c_k T_{t_k}(u) + (1 - c_k)u.$$

Given that $\mathcal{T}$ is a pointwise Lipschitzian semigroup, it is easy to observe that:

$$\|H_k(u) - H_k(v)\| \leq a_{t_k}(u)\|u - v\|.$$

This, along with the summability condition (ii') specified in Definition 4.4, implies that $\{H_k\} \in \mathcal{A}_c(C)$. It is evident that $H_k(x_k) = x_{k+1}$, which shows that the sequence $\{x_k\}$ is generated by $\{H_k\}$. Since

$$w_1, w_2 \in F(\mathcal{F}) \subset \bigcap_{k=1}^{\infty} F(H_k),$$

the equality (4.13) follows directly from Lemma 3.8. $\square$

The proof of the following convergence theorem largely mirrors the approach taken in the proof of Theorem 4.1. However, we substitute The Demiclosedness Principle 1 with The Demiclosedness Principle 2, and replace the argument utilising the Opial inequality (4.12) with the equality (4.13), as established in Lemma 4.4. To ensure completeness and uphold the independence of the results presented in this section, we will provide the full proof here.

Theorem 4.6 *Let X be a uniformly convex and uniformly smooth Banach space X. Let C be a bounded, closed, and convex subset of a X, and let $\mathcal{T} \in \mathcal{APNS}(C)$. Let A be a generating set for J. Assume that $gKM_{UB}(\mathcal{T}, \{c_k\}, \{t_k\})$ is a well defined Krasnosel'skii-Mann iteration process. If the sequence $\{x_k\}$ generated by $gKM_{UB}(\mathcal{T}, \{c_k\}, \{t_k\})$ is an approximate fixed point sequence for every operator T_s, where $s \in A$, then there exists a common fixed point $w \in F(\mathcal{T})$ such that $x_k \rightharpoonup w$.*

4.3 Weak Convergence of Generalised Krasnosel'skii-Mann Iteration ...

Proof By Theorem 3.1, $\{x_k\}$ is an approximate fixed point sequence for any $s \in J$. Consider $y, z \in C$, which are two weak cluster points of the sequence $\{x_k\}$. Then, there exist two subsequences $\{y_k\}$ and $\{z_k\}$ of $\{x_k\}$ such that $y_k \rightharpoonup y$ and $z_k \rightharpoonup z$. Since $\{x_k\}$ is an approximate fixed point sequence for T_s for each $s \in J$, it follows from the Demiclosedness Principle 2 (Theorem 3.5) that $T_s(y) = y$ and $T_s(z) = z$ for all $s \in J$. This means that $y, z \in F(\mathcal{T})$. Therefore, by Lemma 4.4, we obtain the identity

$$\|y - z\|^2 = \langle y - z, J(y - z) \rangle = 0,$$

which implies that $y = z$. We conclude that the sequence $\{x_k\}$ has at most one weak cluster point. Given that C is weakly sequentially compact, it follows that $\{x_k\}$ has exactly one weak cluster point, denoted as $w \in C$. Thus, $x_k \rightharpoonup w$. By applying the Demiclosedness Principle 2 once more, we arrive at the conclusion that $T_s(w) = w$ for any $s \in J$, as claimed. □

The next result is a uniformly smooth analogue of the Theorem 4.2, which assumed the Opial property.

Theorem 4.7 *Let X be a uniformly convex and uniformly smooth Banach space X, and let C be a bounded, closed, and convex subset of X. Consider the semigroup $\mathcal{T} \in \mathcal{APNS}(C)$ with J being generated by a finite set $A = \{\alpha_1, \alpha_2, \alpha_3 ..., \alpha_M\}$ where each α_i is a natural number. Assume that $gKM_{UB}(\mathcal{T}, \{c_k\}, \{t_k\})$ is a well defined Krasnosel'skii-Mann iteration process. Assume that for every natural $m \leq M$, there exists a strictly increasing, quasi-periodic sequence of natural numbers $\{j_k(m)\}$, with a quasi-period p_m, such that for every $k \in \mathbb{N}$, $t_{j_{k+1}(m)} = \alpha_m + t_{j_k(m)}$. Then, the sequence $\{x_k\}$, generated by $gKM_{UB}(\mathcal{T}, \{c_k\}, \{t_k\})$, is an approximate fixed point sequence for every T_s, where $s \in J$. Furthermore, $\{x_k\}$ converges weakly to a common fixed point $w \in F(\mathcal{T})$.*

Proof The proof follows the same approach as that of Theorem 4.2 in the case of the Opial property, with Theorem 4.6 now taking the role of Theorem 4.1. Note also that Lemmas 4.2 and 4.3, needed in the proof, require X to be uniformly convex only. □

Using similar reasoning, it can be concluded that the following analogues of Theorems 4.3, 4.4, and 4.5 remain valid in uniformly convex and uniformly smooth Banach spaces:

Theorem 4.8 *Let X be a uniformly convex and uniformly smooth Banach X, and let C be a bounded, closed, and convex subset of X. Suppose that $T : C \to C$ is nonexpansive and consider the Krasnosel'skii-Mann process defined by the formula*

$$x_{k+1} = c_k T(x_k) + (1 - c_k) x_k.$$

Then, the sequence $\{x_k\}$ generated by this process is an approximate fixed point sequence for T. Furthermore, $\{x_k\}$ converges weakly to a fixed point of T.

Theorem 4.9 *Let X be a uniformly convex and uniformly smooth Banach space X. Let C be a bounded, closed, and convex subset of X. Let $\mathcal{T} \in \mathcal{APNS}(C)$ be a semigroup with $J = [0, +\infty)$. Assume that $\mathcal{T}$ is an equicontinuous semigroup. Let $\{x_k\}$ be the sequence generated by a well defined generalised Krasnosel'skii-Mann iteration process $gKM_{UB}(\mathcal{T}, \{c_k\}, \{t_k\})$. Assume that $B \subset A = [0, 1)$ is dense in A. If to every $s \in B$ there exists a strictly increasing sequence of natural numbers $\{j_k\}$ such that*

(a) $t_{j_{k+1}} - t_{j_k} \to s$ as $k \to \infty$,
(b) $\|x_k - x_{j_k}\| \to 0$ as $k \to \infty$,

then the sequence $\{x_k\}$ converges weakly to a common fixed point $w \in F(\mathcal{T})$.

Theorem 4.10 *Let X be a uniformly convex and uniformly smooth Banach space X. Let C be a bounded, closed, and convex subset of X. Let $\mathcal{T} \in \mathcal{APNS}(C)$ be equicontinuous and let $J = [0, +\infty)$. Let $\{x_k\}$ be generated by a well defined Krasnosel'skii-Mann iteration process $gKM_{UB}(\mathcal{T}, \{c_k\}, \{t_k\})$. Let $s_1, s_2 \in J$ be such that the ratio s_1/s_2 is irrational and that there exist two strictly increasing sequence of natural numbers $\{i_k\}$ and $\{j_k\}$ satisfying*

$$\lim_{k \to \infty} \|x_k - x_{i_k}\| = 0,$$

$$\lim_{k \to \infty} \|x_k - x_{j_k}\| = 0,$$

$$\lim_{k \to \infty} \|T_{e_k}(x_{j_k}) - x_{j_k}\| = 0,$$

$$\lim_{k \to \infty} \|T_{f_k}(x_{j_k}) - x_{j_k}\| = 0,$$

where $e_k = |t_{i_{k+1}} - t_{i_k} - s_1|$ and $f_k = |t_{j_{k+1}} - t_{j_k} - s_2|$, then the sequence $\{x_k\}$ converges weakly to a common fixed point $w \in F(\mathcal{T})$.

4.4 Strong Convergence of Generalised Krasnosel'skii-Mann Iteration Processes

In the previous sections, we observed that the compactness of the set C in the weak topology was crucial for demonstrating the weak convergence of the well defined generalised Krasnosel'skii-Mann process to a common fixed point of a semigroup. It is not surprising that compactness in the norm topology will be essential for achieving the strong convergence of these processes, as will be shown in this section.

As in previous sections, we begin our consideration by examining the discrete case.

Theorem 4.11 *Let X be a uniformly convex Banach space X, and let C be a compact convex subset of X. Assume that the semigroup $\mathcal{T} \in \mathcal{APNS}(C)$ is parametrised by*

4.4 Strong Convergence of Generalised Krasnosel'skii-Mann Iteration Processes

J generated by a finite set $A = \{\alpha_1, \alpha_2, \alpha_3 ..., \alpha_M\}$, where each α_i is a natural number. Assume that $gKM(\mathcal{T}, \{c_k\}, \{t_k\})$ is a well defined Krasnosel'skii-Mann iteration process. Assume that for every natural number $m \leq M$, there exists a strictly increasing, quasi-periodic sequence of natural numbers $\{j_k(m)\}$, with a quasi-period p_m, such that

$$t_{j_{k+1}(m)} = \alpha_m + t_{j_k(m)}$$

holds for every $k \in \mathbb{N}$. Then, the sequence $\{x_k\}$, generated by $gKM(\mathcal{T}, \{c_k\}, \{t_k\})$, is an approximate fixed point sequence for every T_s, where $s \in J$. Furthermore, $\{x_k\}$ converges strongly to a common fixed point $w \in F(\mathcal{T})$.

Proof To apply Lemma 4.3, note that the assumptions (4.6) and (4.7) of Lemma 4.3 are already satisfied since $t_{j_{k+1}(m)} - t_{j_k(m)} - \alpha_m = 0$. To prove (4.5) of the same lemma, observe that by the quasi-periodicity of $\{j_k(m)\}$, to every positive integer k there exists $j_k(m)$ such that $|k - j_k(m)| \leq p_m$. Assume that $k - p_m \leq j_k(m) \leq k$ (the proof for the other case is identical). Fix $\varepsilon > 0$. According to Lemma 4.2, we have that $\|x_{k+1} - x_k\| < \frac{\varepsilon}{p_m}$ for k sufficiently large. Hence for k sufficiently large, we obtain the inequality

$$\|x_k - x_{j_k}\| \leq \|x_k - x_{k-1}\| + ... + \|x_{j_k(m)+1} - x_{j_k(m)}\| \leq p_m \frac{\varepsilon}{p_m} = \varepsilon.$$

This proves that (4.5) is also satisfied. Therefore, it follows from Lemma 4.3 that $\{x_k\}$ is an approximate fixed point sequence for T_s for every $s \in J$.

Let us fix temporarily a number $s \in J$. By compactness of C, there exists a subsequence $\{x_{p_k}\}$ of $\{x_k\}$ and element $x \in C$ such that

$$\|T_s(x_{p_k}) - x\| \to 0 \text{ as } k \to \infty,$$

which, together with the fact that $\{x_k\}$ is an approximate fixed point sequence, implies that

$$\lim_{k \to \infty} \|x_{p_k} - x\| \leq \lim_{k \to \infty} \|x_{p_k} - T_s(x_{p_k})\| + \lim_{k \to \infty} \|T_s(x_{p_k}) - x\| = 0. \quad (4.14)$$

Finally, we conclude that for every $t \in J$,

$$\|T_t(x) - x\| \leq \|T_t(x) - T_t(x_{p_k})\| + \|T_t(x_{p_k}) - x_{p_k}\| + \|x_{p_k} - x\|$$
$$\leq a_t(x_{p_k})\|x_{p_k} - x\| + \|T_t(x_{p_k}) - x_{p_k}\| + \|x_{p_k} - x\|,$$

the right-hand side of which tends to zero as $k \to \infty$, because $a_t(x_{p_k}) \leq M_t < +\infty$ by the regularity of $\{x_k\}$, $\|x_{p_k} - x\| \to 0$, and $\|T_t(x_{p_k}) - x_{p_k}\| \to 0$. Therefore, $T_t(x) = x$ for every $t \in J$. It follows from Lemma 4.1, that $\lim_{k \to \infty} \|x_k - x\|$ exists. This, along with (4.14), implies that $\lim_{k \to \infty} \|x_k - x\| = 0$. The proof is complete. □

It is unsurprising that when C is a compact set, the semigroup continuity and the pointwise boundedness of each a_t imply its equicontinuity, as shown by the following result.

Lemma 4.5 *Let $\mathcal{T}$ be a continuous semigroup. Assume that $\sup\{a_t(x) : t \in J\} < +\infty$ for each $x \in C$. If $C \subset X$ is compact, then $\mathcal{T}$ is equicontinuous.*

Proof Assume to the contrary that this is not the case. Then, there exist $\{t_k\}$ a sequence of numbers from J with $t_k \to 0$, a sequence $\{y_k\}$ of elements in C, and a number $\eta > 0$ such that the inequality

$$\|T_{t_k}(y_k) - y_k\| > \eta > 0$$

holds for every $k \in \mathbb{N}$. Using the compactness of C and passing to a subsequence of $\{y_k\}$, if necessary, we can assume that there exists $w \in C$ such that $\|y_k - w\| \to 0$ as $k \to \infty$. Consider the inequalities

$$0 < \eta \leq \limsup_{k \to \infty} \|T_{t_k}(y_k) - y_k\|$$
$$\leq \limsup_{k \to \infty} \left(\|T_{t_k}(y_k) - T_{t_k}(w)\| + \|T_{t_k}(w) - w\| + \|w - y_k\| \right)$$
$$\leq \limsup_{k \to \infty} \left(a_{t_k}(w)\|y_k - w\| + \|T_{t_k}(w) - w\| + \|w - y_k\| \right)$$
$$\leq \sup_{k \in \mathbb{N}} a_{t_k}(w) \lim_{k \to \infty} \|y_k - w\| + \lim_{k \to \infty} \|T_{t_k}(w) - w\| + \lim_{k \to \infty} \|w - y_k\| = 0.$$

This holds because $\sup_{k \in \mathbb{N}} a_{t_k}(w) < \infty$, the function $J \ni t \mapsto T_t(w)$ is continuous, $T_0(w) = w$, and $\lim_{k \to \infty} \|w - y_k\| = 0$. The contradiction concludes the proof. □

Theorem 4.12 *Let C be a compact, convex subset of a uniformly convex Banach space X. Let $\mathcal{T} \in \mathcal{APNS}(C)$ and $B \subset \overline{B} = A \subset J$ where $A = [0, 1)$ and $J = [0, +\infty)$. Assume that $\sup\{a_t(x) : t \in J\} < \infty$ for every $x \in C$. Let $\{x_k\}$ be a sequence generated by a well defined generalised Krasnosel'skii-Mann iteration process $gKM(\mathcal{T}, \{c_k\}, \{t_k\})$. If to every $s \in B$ there exists a strictly increasing sequence of natural numbers $\{j_k\}$ such that*

(a) $t_{j_{k+1}} - t_{j_k} \to s$ as $k \to \infty$,
(b) $\|x_k - x_{j_k}\| \to 0$ as $k \to \infty$,

then the sequence $\{x_k\}$ converges strongly to a common fixed point $x \in F(\mathcal{T})$.

Proof We apply Lemma 4.3 for the parameter set B. Note that condition (4.5) of Lemma 4.3 is assumed. According to Lemma 4.5, the semigroup $\mathcal{T}$ is equicontinuous, which implies that the assumption (4.7) of Lemma 4.3 is also satisfied. Consequently, it follows from Lemma 4.3 that $\{x_k\}$ is an approximate fixed point sequence for all T_t where $t \in B$. According to Lemma 3.2, the sequence $\{x_k\}$ is an approximate fixed point sequence for any T_t, where $t \in A$. Since A is a generating set for J, it follows from Theorem 3.1 that

4.5 Special Cases

$$\|T_s(x_k) - x_k\| \to 0 \text{ as } k \to \infty \tag{4.15}$$

for every $s \in J$. Let us fix temporarily a number $s \in J$. Since C is a compact set, there exists a subsequence $\{x_{p_k}\}$ of $\{x_k\}$ and element $x \in C$ such that

$$\|T_s(x_{p_k}) - x\| \to 0 \text{ as } k \to \infty. \tag{4.16}$$

Observe that the right-hand of the inequality

$$\|x_{p_k} - x\| \le \|x_{p_k} - T_s(x_{p_k})\| + \|T_s(x_{p_k}) - x\|$$

tends to zero as $k \to \infty$, because of (4.15) and (4.16). Thus,

$$\lim_{k\to\infty} \|x_{p_k} - x\| = 0. \tag{4.17}$$

Finally, for every $t \in J$,

$$\|T_t(x) - x\| \le \|T_t(x) - T_t(x_{p_k})\| + \|T_t(x_{p_k}) - x_{p_k}\| + \|x_{p_k} - x\|$$
$$\le a_t(x_{p_k})\|x_{p_k} - x\| + \|T_t(x_{p_k}) - x_{p_k}\| + \|x_{p_k} - x\|,$$

the right-hand side of which tends to zero as $k \to \infty$, because $a_t(x_{p_k}) \le M_t < +\infty$ by the regularity of $\{x_k\}$, $\|x_{p_k} - x\| \to 0$ by (4.17), and $\|T_t(x_{p_k}) - x_{p_k}\| \to 0$ by (4.15). Therefore, $T_t(x) = x$ for every $t \in J$, which means that x is a common fixed point for the semigroup $\mathcal{T}$. It follows from Lemma 4.1, that $\lim_{k\to\infty} \|x_k - x\|$ exists. This, along with (4.17), implies that $\lim_{k\to\infty} \|x_k - x\| = 0$. The proof is thereby complete. □

4.5 Special Cases

The question arises as to whether a significant simplification can be achieved by reducing the common fixed point problem for the entire $\mathcal{T}$ to a problem of finding a fixed point of just one suitable chosen mapping U. Our next result demonstrates that this simplification is possible for pointwise eventually nonexpansive semigroups, and, consequently, also for nonexpansive semigroups, where $J = [0, +\infty)$.

Theorem 4.13 *Let C be a bounded, closed, and convex subset of a uniformly convex Banach space X. Let $\mathcal{T}$ be a pointwise eventually nonexpansive semigroup with the parameter set $J = [0, +\infty)$. Let $s_1, s_2 \in J$ be such that the ratio s_1/s_2 is an irrational number. Then there exists an $n_0 \in \mathbb{N}$ such that for every natural number $n \ge n_0$,*

$$F(\mathcal{T}) = F(\lambda T_{ns_1} + (1-\lambda)T_{ns_2})$$

holds for every $\lambda \in (0, 1)$.

Proof Since $\mathcal{T}$ as a pointwise eventually nonexpansive semigroup is also an asymptotic pointwise nonexpansive semigroup, it follows from Theorem 2.3 that $F(\mathcal{T})$ is nonempty and consequently that there exists an element $w \in F(T_{s_1}) \cap F(T_{s_2})$. From the definition of pointwise eventually nonexpansive semigroups, it follows that we can choose a number $n_0 \in \mathbb{N}$ such that $a_{ns_1}(w) = a_{ns_2}(w) = 1$ holds for any natural number $n \geq n_0$. Note that $w \in F(T_{ns_1}) \cap F(T_{ns_2})$. Fix an arbitrary $\lambda \in (0, 1)$ and define a mapping V by the formula

$$V(x) = \lambda T_{ns_1}(x) + (1 - \lambda) T_{ns_2}(x) \tag{4.18}$$

for every $x \in C$. We claim that

$$F(V) = F(T_{ns_1}) \cap F(T_{ns_1}). \tag{4.19}$$

Since it is obvious that $F(V) \supset F(T_{ns_1}) \cap F(T_{ns_2})$, we need only to prove the converse inclusion. Let us fix an arbitrary element $x \in F(V)$ and calculate

$$\begin{aligned}
\|x - w\| &= \|\lambda T_{ns_1}(x) + (1 - \lambda) T_{ns_2}(x) - w\| \\
&\leq \lambda \|T_{ns_1}(x) - w\| + (1 - \lambda) \|T_{ns_2}(x) - w\| \\
&= \lambda \|T_{ns_1}(x) - T_{ns_1}(w)\| + (1 - \lambda) \|T_{ns_2}(x) - T_{ns_2}(w)\| \\
&\leq \lambda a_{ns_1}(w) \|x - w\| + (1 - \lambda) a_{ns_2}(w) \|x - w\| \\
&= \lambda \|x - w\| + (1 - \lambda) \|x - w\| \\
&= \|x - w\|.
\end{aligned} \tag{4.20}$$

Since $0 \leq \|T_{ns_i}(x) - w\| \leq \|x - w\|$ it follows from (4.20) that

$$\|T_{ns_1}(x) - w\| = \|T_{ns_2}(x) - w\| = \|x - w\|.$$

Note that X is strictly convex, as it is uniformly convex. Therefore, we infer from the equality

$$\|x - w\| = \|\lambda T_{ns_1}(x) + (1 - \lambda) T_{ns_2}(x) - w\| = \|T_{ns_1}(x) - w\| = \|T_{ns_2}(x) - w\|$$

that

$$x = \lambda T_{ns_1}(x) + (1 - \lambda) T_{ns_2}(x) = T_{ns_1}(x) = T_{ns_2}(x),$$

and consequently that $x \in F(T_{ns_1}) \cap F(T_{ns_1})$, as claimed. From Proposition 3.2, it follows that $F(\mathcal{T}) = F(T_{ns_1}) \cap F(T_{ns_2})$. Together with (4.19), this leads to the conclusion that

$$F(\mathcal{T}) = F(\lambda T_{ns_1} + (1 - \lambda) T_{ns_2}),$$

thereby completing the proof. □

4.5 Special Cases

For eventually nonexpansive semigroups, the construction process can be simplified even more, as explained in the following result.

Theorem 4.14 *Let C be a bounded, closed, and convex subset of a uniformly convex Banach space X, which either possess the Opial property or is uniformly smooth. Let $\mathcal{T}$ be an eventually nonexpansive semigroup with the parameter set $J = [0, +\infty)$. Let $s_1, s_2 \in J$ be such that the ratio s_1/s_2 is an irrational number. Fix any $\kappa_1, \kappa_2 > 0$ with $\kappa_1 + \kappa_2 < 1$. Then, there is a natural number $n \in \mathbb{N}$ such that the sequence $\{x_k\}$ in C defined by $x_1 \in C$ and*

$$x_{k+1} = \kappa_1 T_{ns_1}(x_k) + \kappa_2 T_{ns_2}(x_k) + (1 - \kappa_1 - \kappa_2)x_k, \tag{4.21}$$

converges in the weak topology to a common fixed point of $\mathcal{T}$.

Proof Let $t_0 \in J$ be such that for all $t \in J$ with $t \geq t_0$, the condition $\alpha_t(x) = 1$ holds for every $x \in C$, existing by Definition 1.9. Since $\mathcal{T}$ is also a pointwise eventually nonexpansive semigroup as it is eventually nonexpansive, it follows from Theorem 4.13 that there exists an $n_0 \in \mathbb{N}$ such that for every natural number $n \geq n_0$ the following equality:

$$F(\mathcal{T}) = F(\lambda T_{ns_1} + (1 - \lambda)T_{ns_2}) \tag{4.22}$$

holds for every $\lambda \in (0, 1)$. Let us fix a natural number $n \geq n_0$ such that both $ns_1 \geq t_0$ and $ns_2 \geq t_0$. Define $U : C \to C$ by

$$U(x) = \frac{\kappa_1}{\kappa_1 + \kappa_2} T_{ns_1}(x) + \frac{\kappa_2}{\kappa_1 + \kappa_2} T_{ns_2}(x), \tag{4.23}$$

and observe that U is a nonexpansive mapping. By applying Theorem 4.3 (resp., Theorem 4.8), the sequence $\{x_k\}$, defined by $x_1 \in C$ and

$$x_{k+1} = c_k U(x_k) + (1 - c_k)x_k,$$

where $c_k = \kappa_1 + \kappa_2$, converges weakly to a fixed point w of U. Note that, after substituting (4.23) into (4.22), we obtain the formula (4.21) for x_{k+1}. It follows from the equality (4.22), applied with

$$\lambda = \frac{\kappa_1}{\kappa_1 + \kappa_2},$$

that this w is a common fixed point for the semigroup $\mathcal{T}$, thereby completing the proof. □

Given that it is straightforward to find $s_1, s_2 \in [0, +\infty)$ such that the ratio s_1/s_2 is an irrational number (take, for example, $s_1 = 1$ and $s_2 = \sqrt{2}$), and considering the abundance of well-developed computational implementations of the algorithms (4.21) for finding fixed points of a single nonexpansive operator, the conclusion of our Theorem 4.14 holds significant practical value.

4.6 Notes

The Krasnosel'skii-Mann algorithm is particularly significant due to its inherent simplicity, and its versatility in reducing complex scenarios to a single operator implementation, as we have observed. Moreover, it serves as a fundamental building block for more advanced multi-step iteration processes. The next chapter will elaborate on this with an example of the two-step Ishikawa process.

As already noted, the origins of the method described in this chapter can be traced back to the 1950s works by Mann [82] and Krasnosel'skii [78]. There is a huge body of research related to this iteration method and it is out of scope of this text to provide a comprehensive study of this process. We point the reader's attention to the following works by Xu [107, 108] and by Xu and Tan [105]. The latest comprehensive review of this method can be found in the book by Zaslavski [113].

The results of Sects. 4.2 and 4.4 are loosely based on the 2013 paper by Kozlowski and Sims [77] and follow a long list of earlier studies. We refer the reader to [77] for a more detailed historical background. The uniformly smooth case addressed in Sect. 4.3 follows closely the 2012 paper by Kozlowski [62]. The ideas behind the methods used in Sect. 4.5 have their origin in papers by Suzuki [99–101] and other papers by this author cited therein.

Chapter 5
Generalised Ishikawa Iteration Processes

Abstract In this chapter we consider the two-step Ishikawa iteration process which is a generalisation of the one-step Krasnosel'skii-Mann process. The Ishikawa iteration process provides more flexibility in defining the algorithm parameters which is important from the numerical implementation perspective.

5.1 Preliminaries

Definition 5.1 Let $\mathcal{T} \in \mathcal{APNS}(C)$. Let $\{t_k\}$ be a sequence of elements in J and let $\{c_k\}$ $\{d_k\}$ be two sequences of numbers from $(0, 1)$. The generalised Ishikawa iteration process $gI(\mathcal{T}, \{c_k\}, \{d_k\}, \{t_k\})$ generated by the semigroup $\mathcal{T}$, the sequences $\{c_k\}$, $\{d_k\}$ and $\{t_k\}$, is defined by the iterative formula

$$x_{k+1} = c_k T_{t_k}(d_k T_{t_k}(x_k) + (1 - d_k)x_k) + (1 - c_k)x_k,$$

where $x_1 \in C$ is chosen arbitrarily, and

(i) $\{c_k\}$ is bounded away from 0 and 1, and $\{d_k\}$ is bounded away from 1,
(ii) $\sum_{k=1}^{\infty} b_{t_k}(x) < \infty$ for every $x \in C$.

Definition 5.2 We say that a generalised Ishikawa iteration process is well-defined if the sequence $\{x_k\}$, generated by it, is regular with respect to the semigroup $\mathcal{T}$, and for any $s \in J$,

$$\limsup_{k \to \infty} a_{s+t_k}(x_k) = 1.$$

The following result represents an analogue of Lemma 4.1, originally established for the Krasnosel'skii-Mann process, now applied to the Ishikawa process.

Lemma 5.1 *Let C be a bounded, closed, and convex subset of a Banach space X. Let $\mathcal{T} \in \mathcal{APNS}(C)$, $w \in F(\mathcal{T})$, and let $gI(\mathcal{T}, \{c_k\}, \{d_k\}, \{t_k\})$ be a generalised Ishikawa process. Then, there exists a number $r \in \mathbb{R}$ such that $\lim_{k \to \infty} \|x_k - w\| = r$.*

Proof Define $G_k : C \to C$ by

$$G_k(x) = c_k T_{t_k}(d_k T_{t_k}(x) + (1-d_k)x) + (1-c_k)x.$$

Clearly, $x_{k+1} = G_k(x_k)$. Note also that $w \in F(\mathcal{T}) \subset \bigcap_{k=1}^{\infty} F(G_k)$. For the simplicity sake, let us use the notation

$$S_k(x) = d_k T_{t_k}(x) + (1-d_k)x,$$

and

$$H_k(x) = T_{t_k}(S_k(x)).$$

Thus, we have the representation

$$G_k(x) = c_k H_k(x) + (1-c_k)x.$$

Furthermore, a straight calculation shows that

$$\|H_k(x) - H_k(y)\| \leq a_{t_k}(S_k(x))(d_k(a_{t_k}(x) + (1-d_k))\|x-y\| \quad (5.1)$$

and

$$\|G_k(x) - G_k(y)\| \leq c_k \|H_k(x) - H_k(y)\| + (1-c_k)\|x-y\|. \quad (5.2)$$

By combining (5.1) and (5.2), we conclude that each G_k satisfies the inequality

$$\|G_k(x) - G_k(y)\| \leq A_k(x)\|x-y\|,$$

where

$$A_k(x) = 1 + c_k a_{t_k}(d_k T_{t_k}(x) + (1-d_k)x)(1 + d_k a_{t_k}(x) - d_k) - c_k. \quad (5.3)$$

Note that $A_k(x) \geq 1$, which follows directly from the fact that $a_{t_k}(z) \geq 1$ for any $z \in C$. Since $\lim_{k \to \infty} a_{t_k}(w) = 1$, it follows that there exists an $M > 0$ such that $a_{t_k}(w) \leq M$ for all $k \in \mathbb{N}$. By using (5.3) and considering that $w \in F(\mathcal{T})$, we obtain the estimation

$$\begin{aligned} B_k(w) &= A_k(w) - 1 \quad (5.4)\\ &= c_k a_{t_k}(w)(1 + d_k a_{t_k}(w)) - c_k d_k a_{t_k}(w) - c_k \\ &= c_k(1 + d_k a_{t_k}(w))(a_{t_k}(w) - 1) \\ &\leq (1 + a_{t_k}(w))b_{t_k}(w) \leq (1+M)b_{t_k}(w). \end{aligned}$$

Since $\sum_{k=1}^{\infty} b_{t_k}(w) < \infty$, it follows from (5.4) that $\sum_{k=1}^{\infty} B_k(w) < \infty$. Consequently, the sequence $\{G_k\} \in \mathcal{A}(C)$ is an asymptotic nonexpansive sequence that

5.1 Preliminaries

satisfies the assumptions of Lemma 3.5. This implies the existence of an $r \in \mathbb{R}$ such that $\lim_{k\to\infty} \|x_k - w\| = r$. □

Our next result plays a similar role in the context of the Ishikawa process as Lemma 4.2 did for the generalised Krasnosel'skii-Mann process.

Lemma 5.2 *Let C be a bounded, closed, and convex subset of a uniformly convex Banach space X. Let $\mathcal{T} \in \mathcal{APNS}(C)$. Let $gI(\mathcal{F}, \{c_k\}, \{d_k\}, \{t_k\})$ be a well-defined generalised Ishikawa iteration process. Then,*

$$\lim_{k\to\infty} \|T_{t_k}(x_k) - x_k\| = 0 \tag{5.5}$$

and

$$\lim_{k\to\infty} \|x_{k+1} - x_k\| = 0. \tag{5.6}$$

Proof By Theorem 2.3, $F(\mathcal{T}) \neq \emptyset$. Choose an arbitrary fixed point $w \in F(\mathcal{T})$. According to Lemma 5.1, the limit $\lim_{k\to\infty} \|x_k - w\|$ exists. Let us denote it by r. Let us define

$$y_k = d_k T_{t_k}(x_k) + (1 - d_k)x_k.$$

Since $w \in F(\mathcal{T})$, $\mathcal{T} \in \mathcal{APNS}(C)$ implying that $\lim_{k\to\infty} a_{t_k}(w) = 1$, and the limit $\lim_{k\to\infty} \|x_k - w\| = r$, we conclude that

$$\begin{aligned}
\limsup_{k\to\infty} \|T_{t_k}(y_k) - w\| &= \limsup_{k\to\infty} \|T_{t_k}(y_k) - T_{t_k}(w)\| \\
&\leq \limsup_{k\to\infty} a_{t_k}(w)\|y_k - w\| \\
&= \limsup_{k\to\infty} a_{t_k}(w)\|d_k T_{t_k}(x_k) + (1-d_k)x_k - w\| \\
&\leq \limsup_{k\to\infty} \left(d_k a_{t_k}(w)\|T_{t_k}(x_k) - w\| + (1-d_k)a_{t_k}(w)\|x_k - w\| \right) \\
&\leq \lim_{k\to\infty} \left(d_k a_{t_k}^2(w)\|x_k - w\| + (1-d_k)a_{t_k}(w)\|x_k - w\| \right) \leq r.
\end{aligned}$$

Note that

$$\lim_{k\to\infty} \|d_k(T_{t_k}(y_k) - w) + (1-d_k)(x_k - w)\| = \lim_{k\to\infty} \|d_k T_{t_k}(y_k) + (1-d_k)x_k - w\|$$
$$= \lim_{k\to\infty} \|x_{k+1} - w\| = r.$$

Applying Lemma 3.3 with $u_k = T_{t_k}(y_k) - w$ and $v_k = x_k - w$, we obtain the equality

$$\lim_{k\to\infty} \|T_{t_k}(y_k) - x_k\| = 0. \tag{5.7}$$

This fact, combined with the construction formulas for x_{k+1} and y_k, gives us (5.6). Since

$$\|T_{t_k}(x_k) - x_k\| \leq \|T_{t_k}(x_k) - T_{t_k}(y_k)\| + \|T_{t_k}(y_k) - x_k\|$$
$$\leq a_{t_k}(x_k)\|x_k - y_k\| + \|T_{t_k}(y_k) - x_k\|$$
$$= d_k a_{t_k}(x_k)\|T_{t_k}(x_k) - x_k\| + \|T_{t_k}(y_k) - x_k\|,$$

it follows that

$$\|T_{t_k}(x_k) - x_k\| \leq (1 - d_k a_{t_k}(x_k))^{-1}\|T_{t_k}(y_k) - x_k\|. \tag{5.8}$$

The right-hand side of inequality (5.8) tends to zero due to the following reasons: (a) $\|T_{t_k}(y_k) - x_k\| \to 0$ by (5.7), (b) $\lim_{k\to\infty} a_{t_k}(x_k) = 1$ because the Ishikawa process is assumed to be well-defined, (c) the sequence $\{d_k\}$ consisting of numbers from the interval $(0, 1)$ is bounded away from 1. Therefore, we conclude that the limit $\lim_{k\to\infty} \|T_{t_k}(x_k) - x_k\| = 0$, as claimed in (5.5). This completes the proof. □

We need the following technical result being the Ishikawa version of Lemma 4.3. Although the proof follows a similar structure, with Lemma 4.2 replaced by Lemma 5.2, we present a concise outline of it for the purpose of completeness.

Lemma 5.3 *Let C be a bounded, closed, and convex subset of a uniformly convex Banach space X. Let $\mathcal{T} \in \mathcal{APNS}(C)$. Let the generalised Ishikawa process $gI(\mathcal{T}, \{c_k\}, \{d_k\}, \{t_k\})$ be well-defined. Let $A \subset J$ be such that to every $s \in A$ there exists a strictly increasing sequence of natural numbers $\{j_k\}$ satisfying the following conditions:*

$$\lim_{k\to\infty} \|x_k - x_{j_k}\| = 0, \tag{5.9}$$

$$e_k := |t_{j_{k+1}} - t_{j_k} - s| \in J, \text{ for all } k \in \mathbb{N}, \tag{5.10}$$

and

$$\lim_{k\to\infty} \|T_{e_k}(x_{j_k}) - x_{j_k}\| = 0. \tag{5.11}$$

Then, $\{x_k\}$ is an approximate fixed point sequence for all mappings $\{T_{ms}\}$, where $s \in A$ and $m \in \mathbb{N}$, that is

$$\lim_{k\to\infty} \|T_{ms}(x_k) - x_k\| = 0 \tag{5.12}$$

for every $s \in A$ and $m \in \mathbb{N}$. If, in addition, A is a generating set for J, then

$$\lim_{k\to\infty} \|T_t(x_k) - x_k\| = 0$$

for any $t \in J$.

Proof By Lemma 3.4, it suffices to demonstrate equality (5.12) for $m = 1$. Fix an $s \in A$. It follows from the assumption (5.9) along the equality (5.6) from Lemma 5.2 that

$$\|x_{j_k} - x_{j_{k+1}}\| \leq \|x_{j_k} - x_k\| + \|x_k - x_{k+1}\| + \|x_{k+1} - x_{j_{k+1}}\| \to 0.$$

5.2 Weak Convergence of Generalised Ishikawa Iteration Processes ...

By utilising the above equation together with Lemma 5.2 and our assumptions, we observe that the following expression tends to the zero as $k \to \infty$,

$$\begin{aligned}\|x_{j_k} - T_s(x_{j_k})\| &\leq \|x_{j_k} - x_{j_{k+1}}\| + \|x_{j_{k+1}} - T_{t_{j_{k+1}}}(x_{j_{k+1}})\| + \|T_{t_{j_{k+1}}}(x_{j_{k+1}}) - T_{t_{j_{k+1}}}(x_{j_k})\| \\ &\quad + \|T_{t_{j_{k+1}}}(x_{j_k}) - T_{s+t_{j_k}}(x_{j_k})\| + \|T_{s+t_{j_k}}(x_{j_k}) - T_s(x_{j_k})\| \\ &\leq \|x_{j_k} - x_{j_{k+1}}\| + \|x_{j_{k+1}} - T_{t_{j_{k+1}}}(x_{j_{k+1}})\| + a_{t_{j_{k+1}}}(x_{j_{k+1}})\|x_{j_{k+1}} - x_{j_k}\| \\ &\quad + a_{s+t_{j_k}}(x_{j_k})\|T_{e_k}(x_{j_k}) - x_{j_k}\| + \sup_{n \in \mathbb{N}} a_s(x_{j_n})\|T_{t_{j_k}}(x_{j_k}) - x_{j_k}\|.\end{aligned}$$

Using this inequality along the estimation

$$\begin{aligned}\|x_k - T_s(x_k)\| &\leq \|x_k - x_{j_k}\| + \|x_{j_k} - T_{t_{j_k}}(x_{j_k})\| + \|T_{t_{j_k}}(x_{j_k}) - T_{s+t_{j_k}}(x_{j_k})\| \\ &\quad + \|T_{s+t_{j_k}}(x_{j_k}) - T_s(x_{j_k})\| + \|T_s(x_{j_k}) - T_s(x_k)\| \\ &\leq \|x_k - x_{j_k}\| + \|x_{j_k} - T_{t_{j_k}}(x_{j_k})\| + a_{t_{j_k}}(x_{j_k})\|x_{j_k} - T_s(x_{j_k})\| \\ &\quad + a_s(x_{j_k})\|T_{t_{j_k}}(x_{j_k}) - x_{j_k}\| + a_s(x_k)\|x_{j_k} - x_k\|,\end{aligned}$$

we conclude that $\|x_k - T_s(x_k)\| \to 0$ as $k \to \infty$. This means that $\{x_k\}$ is an approximate fixed point sequence for the mapping $\{T_s\}$. If A is a generating set for J, then by Lemma 3.1, $\{x_k\}$ is an approximate fixed point sequence for any $T_s \in \mathcal{T}$. This concludes the proof of the lemma. □

Remark 5.1 Observe that, like in the Krasnosel'skii-Mann case, when $J = [0, +\infty)$, the assumption given in (5.10) of Lemma 5.3 is automatically satisfied. Therefore, in these instances, verification of this assumption is unnecessary.

5.2 Weak Convergence of Generalised Ishikawa Iteration Processes in Banach Spaces with the Opial Property

We are now ready to present the following weak convergence result for the Ishikawa iteration processes. The proof follows a similar structure to that of Theorem 4.1 with Lemma 4.1 replaced by Lemma 5.1. To maintain completeness, we provide a brief sketch of the proof below.

Theorem 5.1 *Let X be a uniformly convex Banach space X with the Opial property. Let C be a bounded, closed, and convex subset of a X. Let $\mathcal{T} \in \mathcal{APNS}(C)$. Assume that $gI(\mathcal{T}, \{c_k\}, \{d_k\}, \{t_k\})$ is a well-defined Ishikawa iteration process. If the sequence $\{x_k\}$ generated by $gI(\mathcal{T}, \{c_k\}, \{d_k\}, \{t_k\})$ is an approximate fixed point sequence for every $s \in A \subset J$, where A is a generating set for J, then $\{x_k\}$ converges weakly to a common fixed point $w \in F(\mathcal{T})$.*

Proof By Theorem 3.1, $\{x_k\}$ is an approximate fixed point sequence for any $s \in J$. Fix any two weak cluster points of the sequence $\{x_k\}$, denoted by y and z. Then, there exist two subsequences $\{y_k\}$ and $\{z_k\}$ of $\{x_k\}$ such that $y_k \rightharpoonup y$ and $z_k \rightharpoonup z$.

Since $\{x_k\}$ is an approximate fixed point sequence for all T_s, it follows from the Demiclosedness Principle 1 (Theorem 3.4) that $T_s(y) = y$ and $T_s(z) = z$ for any $s \in J$, that is, $y, z \in F(\mathcal{T})$. By Lemma 5.1 the limits

$$r_1 = \lim_{k\to\infty} \|x_k - y\|, \quad r_2 = \lim_{k\to\infty} \|x_k - z\|$$

exist. We claim that $y = z$. Assume that this is not the case, then, by the Opial property,

$$r_1 = \liminf_{k\to\infty} \|y_k - y\| < \liminf_{k\to\infty} \|y_k - z\| = r_2$$
$$= \liminf_{k\to\infty} \|z_k - z\| < \liminf_{k\to\infty} \|z_k - y\| = r_1.$$

The contradiction implies $y = z$ which means that the sequence $\{x_k\}$ has at most one weak cluster point. Since C is weakly sequentially compact, we deduce that the sequence $\{x_k\}$ has exactly one weak cluster point $w \in C$, which means that $x_k \rightharpoonup w$. By applying the Demiclosedness Principle again, we conclude that $T_s(w) = w$ for every $s \in J$. □

Similarly, it is easy to modify the proof of Theorems 4.2 and 4.4 to obtain the next two results.

Theorem 5.2 *Let X be a uniformly convex Banach space X with the Opial property, and let C be a bounded, closed, and convex subset of X. Assume that the semigroup $\mathcal{T} \in \mathcal{APNS}(C)$ is parametrised by a subsemigroup J of the additive semigroup $\mathbb{N}_0$, which implies that J is generated by a finite set $A = \{\alpha_1, \alpha_2, \alpha_3..., \alpha_M\}$ where each α_i is a natural number. Assume that $gI(\mathcal{T}, \{c_k\}, \{d_k\}, \{t_k\})$ is a well-defined Ishikawa iteration process. Assume that for every natural number $m \leq M$, there exists a strictly increasing, quasi-periodic sequence of natural numbers $\{j_k(m)\}$, with a quasi-period p_m, such that*

$$t_{j_{k+1}(m)} = \alpha_m + t_{j_k(m)}$$

holds for every $k \in \mathbb{N}$. Then, the sequence $\{x_k\}$, generated by $gI(\mathcal{F}, \{c_k\}, \{d_k\}, \{t_k\})$, is an approximate fixed point sequence for every T_s, where $s \in A$. Furthermore, $\{x_k\}$ converges weakly to a common fixed point $w \in F(\mathcal{T})$.

Theorem 5.3 *Let X be a uniformly convex Banach space X with the Opial property. Let C be a bounded, closed, and convex subset of X. Let $\mathcal{T} \in \mathcal{APNS}(C)$ be a semigroup with $J = [0, +\infty)$. Assume that $\mathcal{T}$ is an equicontinuous semigroup. Let $\{x_k\}$ be the sequence generated by a well defined generalised Ishikawa iteration process $gI(\mathcal{T}, \{c_k\}, \{d_k\}, \{t_k\})$. Assume that $B \subset A = [0, 1)$ is dense in A. If to every $s \in B$ there exists a strictly increasing sequence of natural numbers $\{j_k\}$ such that*

(a) $t_{j_{k+1}} - t_{j_k} \to s$ *as* $k \to \infty$,
(b) $\|x_k - x_{j_k}\| \to 0$ *as* $k \to \infty$,

then the sequence $\{x_k\}$ converges weakly to a common fixed point $w \in F(\mathcal{T})$.

5.3 Weak Convergence of Generalised Ishikawa Iteration Processes in Uniformly Convex and Uniformly Smooth Banach Spaces

Following the approach taken in Sect. 4.3 for the Krasnosel'skii-Mann iteration process, we now assume that the uniformly convex Banach space X is also uniformly smooth. Again, we will narrow our focus in this section to locally uniformly bounded Ishikawa processes, as per the following definition.

Definition 5.3 A generalised Ishikawa iteration process is said to be locally uniformly bounded, and denoted as $gI_{UB}(\mathcal{T}, \{c_k\}, \{d_k\}, \{t_k\})$, if the assumption:

(ii) $\sum_{n=1}^{\infty} b_{t_n}(x) < \infty$ for every $x \in C$.

of Definition 5.1 is replaced by a more general condition:

(ii') For every $x \in C$ there exists a positive constant ε_x and a series of positive constants $\{M_n(x)\}$ such that

$$\sum_{n=1}^{\infty} M_n(x_n) < \infty,$$

and

$$b_{t_n}(y) \leq M_n(x)$$

holds for every $y \in C \cap B(x, \varepsilon_x)$.

To keep the discussion succinct, in the remainder of this section, we will use the notation $gI_{UB}(\mathcal{T}, \{c_k\}, \{d_k\}, \{t_k\})$ to denote a locally uniformly bounded generalised Ishikawa iteration process, even when the term "locally uniformly bounded" is not explicitly stated.

Lemma 5.4 *Let C be a bounded, closed, and convex subset of a uniformly convex and uniformly smooth Banach space X. Let $\mathcal{T} \in \mathcal{APNS}(C)$. Let $\{c_k\}$ be bounded away from 0 and 1, and $\{d_k\}$ be such that $d_k \to 0$. Let $gI_{UB}(\mathcal{T}, \{c_k\}, \{d_k\}, \{t_k\})$ be a well-defined generalised Ishikawa iteration process. Then for every two fixed points $w_1, w_2 \in F(\mathcal{T})$ and for any two weak cluster points y, z of the sequence $\{x_k\}$ generated by $gI_{UB}(\mathcal{T}, \{c_k\}, \{d_k\}, \{t_k\})$,*

$$\langle y - z, J(w_1 - w_2) \rangle = 0. \tag{5.13}$$

Proof Let $w_1, w_2 \in F(\mathcal{T})$. Let us define

$$H_k(u) = c_k T_{t_k}(d_k T_{t_k}(u) + (1-d_k)u) + (1-c_k)u.$$

We need to show that the sequence $\{H_k\} \in \mathcal{A}_c(C)$. Fix any $u, v \in C$ and denote $u_k = d_k T_{t_k}(u) + (1-d_k)u$ and $v_k = d_k T_{t_k}(v) + (1-d_k)v$. Calculate

$$\begin{aligned}
\|H_k(u) - H_k(v)\| &\leq c_k \|T_{t_k}(u_k) - T_{t_k}(v_k)\| + (1-c_k)\|u-v\| \\
&\leq c_k a_{t_k}(u_k)\|u_k - v_k\| + (1-c_k)\|u-v\| \\
&\leq c_k a_{t_k}(u_k)(d_k\|T_{t_k}(u) - T_{t_k}(v)\| + (1-d_k)\|u-v\|) \\
&\quad + (1-c_k)\|u-v\| \\
&\leq c_k a_{t_k}(u_k)(d_k a_{t_k}(u)\|u-v\| + (1-d_k)\|u-v\|) \\
&\quad + (1-c_k)\|u-v\| \\
&\leq c_k a_{t_k}(u_k) a_{t_k}(u)\|u-v\| + (1-c_k)\|u-v\| \\
&\leq a_{t_k}(u_k) a_{t_k}(u)\|u-v\| \\
&= A_k(u)\|u-v\|,
\end{aligned}$$

where $A_k(u) := a_{t_k}(u_k) a_{t_k}(u)$. By the summability condition (ii') from Definition 5.3 there exist $\varepsilon_u > 0$ and a sequence of positive constants $\{M_k(u)\}$ such that $\sum_{k=1}^{\infty} M_k(u) < \infty$ and $b_{t_k}(y) \leq M_k(u)$ holds for every $y \in C \cap B(u, \varepsilon_u)$. Note that

$$\|u_k - u\| = d_k \|T_{t_k}(u) - u\| \leq d_k \, diam(C) \to 0$$

as $k \to \infty$ since $d_k \to 0$. Hence, there exists $k_0 \in \mathbb{N}$ such that $\|u_k - u\| < \varepsilon_u$ for $k > k_0$. This implies that $b_{t_k}(u_k) \leq M_k(u)$ for $k > k_0$. Consequently, $b_{t_k}(u_k) \to 0$, since $\sum_{k=1}^{\infty} M_k(u) < \infty$. Finally, using also the fact that the process is well-defined, we have that $A_k(u) \to 1$ as $k \to \infty$. This implies that the condition (3.20) of Definition 3.6 is satisfied. Therefore, $\{H_k\} \in \mathcal{A}(C)$. We still need to prove that $\{H_k\}$ belongs to $\mathcal{A}_c(C)$, which is a subclass of $\mathcal{A}(C)$; recall Definition 3.7. Let us note that for $k > k_0$,

$$B_k(u) = A_k(u) - 1 = (1 + b_{t_k}(u_k))a_{t_k}(u) - 1 \leq (1 + M_k(u))^2 - 1.$$

Set

$$N_k(u) = (1 + M_k(u))^2 - 1$$

for $k > k_0$ and

$$N_k(u) = m_k(1 + M_k(u)) - 1$$

for $k \leq k_0$, where $m_k = \max\{1 + b_{t_k}(u_k) : k = 1, \ldots, k_0\}$. Clearly, $\sum_{k=1}^{\infty} N_k(u) < \infty$, and $B_k(y) \leq N_k(u)$ for every $y \in C \cap B(u, \varepsilon_u)$. Therefore, by Remark 3.4, $\{H_k\} \in \mathcal{A}_c(C)$.

It is immediate that $H_k(x_k) = x_{k+1}$. Since $w_1, w_2 \in F(\mathcal{T}) \subset \bigcap_{k=1}^{\infty} F(H_k)$, the equality (5.13) follows immediately from Lemma 3.8. This completes the proof. □

The following weak convergence result is an Ishikawa counterpart to the Krasnosel'skii-Mann related Theorem 4.6. The proof follows the similar path.

Theorem 5.4 *Let X be a uniformly convex and uniformly smooth Banach space X. Let C be a bounded, closed, and convex subset of a X. Let $\mathcal{T} \in \mathcal{APNS}(C)$. Assume that $gI_{UB}(\mathcal{T}, \{c_k\}, \{d_k\}, \{t_k\})$ is a well defined Ishikawa iteration process,* and that $d_k \to 0$. If the sequence $\{x_k\}$ generated by $gI_{UB}(\mathcal{T}, \{c_k\}, \{d_k\}, \{t_k\})$ is an approximate fixed point sequence for every $s \in A \subset J$ where A is a generating set for J, then there exists a common fixed point $w \in F(\mathcal{T})$ such that $x_k \rightharpoonup w$.

Proof By Theorem 3.1, $\{x_k\}$ is an approximate fixed point sequence for any $s \in J$. Suppose $y, z \in C$ are two weak cluster points of the sequence $\{x_k\}$. Then, there exist two subsequences $\{y_k\}$ and $\{z_k\}$ of $\{x_k\}$ such that $y_k \rightharpoonup y$ and $z_k \rightharpoonup z$. Since $\{x_k\}$ is an approximate fixed point sequence for T_s for each $s \in J$, it follows from the Demiclosedness Principle 2 (Theorem 3.5) that $T_s(y) = y$ and $T_s(z) = z$ for all $s \in J$. Therefore, $y, z \in F(\mathcal{T})$, which implies by Lemma 5.4 that

$$\|y - z\|^2 = \langle y - z, J(y - z) \rangle = 0.$$

The equality implies that $y = z$. We conclude that the sequence $\{x_k\}$ has at most one weak cluster point. Given that C is weakly sequentially compact, it follows that $\{x_k\}$ has exactly one weak cluster point, denoted as $w \in C$. Thus, $x_k \rightharpoonup w$. By applying the Demiclosedness Principle 2 once more, we arrive at the conclusion that $T_s(w) = w$ for any $s \in J$, as claimed. □

Let us prove the discrete specialisation of the above result, in which we follow the path of Theorems 5.2 and 4.2.

Theorem 5.5 *Let X be a uniformly convex and uniformly smooth Banach space X. Let C be a bounded, closed, and convex subset of X. Assume that the semigroup $\mathcal{T} \in \mathcal{APNS}(C)$ is parametrised by a subsemigroup J of the additive semigroup $\mathbb{N}_0$, which implies that J is generated by a finite set $A = \{\alpha_1, \alpha_2, \alpha_3..., \alpha_M\}$ where each α_i is a natural number. Assume also that $gI(\mathcal{T}, \{c_k\}, \{d_k\}, \{t_k\})$ is a well-defined Ishikawa iteration process, where $d_k \to 0$. Suppose that to every natural number $m \leq M$ there exists a strictly increasing, quasi-periodic sequence of natural numbers $\{j_k(m)\}$, with a quasi-period p_m, such that*

$$t_{j_{k+1}(m)} = \alpha_m + t_{j_k(m)}$$

holds for every $k \in \mathbb{N}$. Then, the sequence $\{x_k\}$, generated by $gI(\mathcal{F}, \{c_k\}, \{d_k\}, \{t_k\})$, is an approximate fixed point sequence for every T_s, where $s \in A$. Furthermore, $\{x_k\}$ converges weakly to a common fixed point $w \in F(\mathcal{T})$.

Proof We will apply Lemma 5.3. Note that the assumptions (5.10) and (5.11) of Lemma 5.3 are trivially satisfied since $t_{j_{k+1}(m)} - t_{j_k(m)} - \alpha_m = 0$. To prove (5.9) of the same lemma, observe that by the quasi-periodicity of $\{j_k(m)\}$, to every positive integer k there exists $j_k(m)$ such that $|k - j_k(m)| \leq p_m$. Assume that $k - p_m \leq j_k(m) \leq k$ (the proof for the other case is identical). Fix $\varepsilon > 0$. Note that by Lemma 5.2, $\|x_{k+1} - x_k\| < \frac{\varepsilon}{p_m}$ for k sufficiently large. Hence for k sufficiently large, we have that

$$\|x_k - x_{j_k}\| \leq \|x_k - x_{k-1}\| + ... + \|x_{j_k(m)+1} - x_{j_k(m)}\| \leq p_m \frac{\varepsilon}{p_m} = \varepsilon.$$

This proves that (5.9) is also satisfied. Therefore, by Lemma 5.3, $\{x_k\}$ is an approximate fixed point sequence for every T_s where $s \in A$. By Theorem 5.4, the sequence $\{x_k\}$ converges weakly to a common fixed point $w \in F(\mathcal{T})$. □

As in the previous cases, the practical importance of our next result is highlighted by the possibility of making the set B countable, which can significantly streamline the computations in case $J = [0, +\infty)$; recall that an interval $A = [0, 1)$ can serve as a convenient choice of a generating set for $J = [0, +\infty)$.

Theorem 5.6 *Let X be a uniformly convex and uniformly smooth Banach space X. Let C be a bounded, closed, and convex subset of X. Let $\mathcal{T} \in \mathcal{APNS}(C)$ be a semigroup with $J = [0, +\infty)$. Assume that $\mathcal{T}$ is equicontinuous. Assume that $B \subset \overline{B} = A$ where A is a generating set for J. Let $\{x_k\}$ be generated by a well defined Ishikawa iteration process $gI_{UB}(\mathcal{T}, \{c_k\}, \{d_k\}, \{t_k\})$, where $d_k \to 0$. If to every $s \in B$ there exists a strictly increasing sequence of natural numbers $\{j_k\}$ such that*

(a) $t_{j_{k+1}} - t_{j_k} \to s$ as $k \to \infty$,
(b) $\|x_k - x_{j_k}\| \to 0$ as $k \to \infty$,

then the sequence $\{x_k\}$ converges weakly to a common fixed point $w \in F(\mathcal{T})$.

Proof We intend to apply Lemma 5.3. To this end, let us note that the condition (5.9) in this lemma is satisfied due to our assumption (b), while (5.11) follows readily from (a) through the equicontinuity of $\mathcal{T}$. Thus, by Lemma 5.3, the sequence $\{x_k\}$ is an approximate fixed point sequence for all mappings $\{T_s\}$ where $s \in B$. Since B is a dense subset of A, it follows from Theorem 3.2 that sequence $\{x_k\}$ is an approximate fixed point sequence for mappings $\{T_s\}$ for all $s \in A$. Since A is a generating set for J, our assertion follows from Theorem 5.4. □

5.4 Strong Convergence of Generalised Ishikawa Iteration Processes

Let us examine the following modifications of the generalised Krasnosel'skii-Mann strong convergence results from Sect. 4.4, adapted to the Ishikawa setting.

Theorem 5.7 *Let X be a uniformly convex Banach space X, and let C be a compact convex subset of X. Assume that the semigroup $\mathcal{T} \in \mathcal{APNS}(C)$ is parametrised by J generated by a finite set $A = \{\alpha_1, \alpha_2, \alpha_3..., \alpha_M\}$, where each α_i is a natural number. Let $gI(\mathcal{T}, \{c_k\}, \{d_k\}, \{t_k\})$ be a well-defined Ishikawa iteration process. Assume that for every natural number $m \leq M$, there exists a strictly increasing, quasi-periodic sequence of natural numbers $\{j_k(m)\}$, with a quasi-period p_m, such that*

$$t_{j_{k+1}(m)} = \alpha_m + t_{j_k(m)}$$

5.4 Strong Convergence of Generalised Ishikawa Iteration Processes

holds for every $k \in \mathbb{N}$. Then, the sequence $\{x_k\}$, generated by $gI(\mathcal{T}, \{c_k\}, \{d_k\}, \{t_k\})$, is an approximate fixed point sequence for every T_s, where $s \in J$. Furthermore, $\{x_k\}$ converges strongly to a common fixed point $w \in F(\mathcal{T})$.

Proof The proof structure closely follows that used in the proof of Theorem 4.11. However, instead of Lemmas 4.1, 4.2, and 4.3, the proof utilises Lemmas 5.1, 5.2, and 5.3, respectively. □

Theorem 5.8 *Let C be a compact, convex subset of a uniformly convex Banach space X. Let $\mathcal{T} \in \mathcal{APNS}(C)$ and $B \subset \overline{B} = A \subset J$ where $A = [0, 1)$ and $J = [0, +\infty)$. Assume that $\sup\{a_t(x) : t \in J\} < \infty$ for every $x \in C$. Let $\{x_k\}$ be a sequence generated by a well-defined generalised Ishikawa iteration process $gI(\mathcal{T}, \{c_k\}, \{d_k\}, \{t_k\})$. If to every $s \in B$ there exists a strictly increasing sequence of natural numbers $\{j_k\}$ such that*

$$\lim_{k \to \infty}(t_{j_{k+1}} - t_{j_k}) = s$$

and

$$\lim_{k \to \infty} \|x_k - x_{j_k}\| = 0, \qquad (5.14)$$

then the sequence $\{x_k\}$ converges strongly to a common fixed point $x \in F(\mathcal{T})$.

Proof We apply Lemma 5.3 with the parameter set B. Note that condition (5.9) of Lemma 5.3 is assumed in (5.14). According to Lemma 4.5, the semigroup $\mathcal{T}$ is equicontinuous, which implies that the assumption (5.11) of Lemma 5.3 is also satisfied. It follows from Lemma 5.3 that $\{x_k\}$ is an approximate fixed point sequence for all T_t where $t \in B$, and by Theorem 3.2, for any $t \in A$. Since A is a generating set for J, it follows from Theorem 3.1 that

$$\|T_s(x_k) - x_k\| \to 0 \text{ as } k \to \infty \qquad (5.15)$$

for every $s \in J$. Let us fix temporarily a number $s \in J$. Since C is a compact set, there exists a subsequence $\{x_{p_k}\}$ of $\{x_k\}$ and element $x \in C$ such that

$$\|T_s(x_{p_k}) - x\| \to 0 \text{ as } k \to \infty. \qquad (5.16)$$

From (5.15) and (5.16) it follows that

$$\lim_{k \to \infty} \|x_{p_k} - x\| \leq \lim_{k \to \infty} \|x_{p_k} - T_s(x_{p_k})\| + \lim_{k \to \infty} \|T_s(x_{p_k}) - x\| = 0. \qquad (5.17)$$

Let $t \in J$. It follows from the regularity of $\{x_k\}$ along with (5.17) and (5.15) that the right-hand side of the following inequality tends to zero as $k \to \infty$:

$$\|T_t(x) - x\| \leq \|T_t(x) - T_t(x_{p_k})\| + \|T_t(x_{p_k}) - x_{p_k}\| + \|x_{p_k} - x\|$$
$$\leq a_t(x_{p_k})\|x_{p_k} - x\| + \|T_t(x_{p_k}) - x_{p_k}\| + \|x_{p_k} - x\|.$$

Thus, $T_t(x) = x$ for every $t \in J$. It follows from Lemma 5.1, that $\lim_{k\to\infty} \|x_k - x\|$ exists. This, along with (5.17), implies that $\lim_{k\to\infty} \|x_k - x\| = 0$. The proof is complete. $\qquad\square$

5.5 Notes

The origins of this method can be traced back to the work of Ishikawa [45]. Like the Kransosel'skii-Mann process, the Ishikawa method has been extensively used in practice and there exists a large body of literature related to this process and its convergence. The results of Sects. 5.2 and 5.4 are loosely based on the 2013 paper by Kozlowski and Sims [77]. The uniformly smooth case addressed in Sect. 5.3 is loosely based on the 2012 paper by Kozlowski [62].

Chapter 6
Implicit Iteration Processes

Abstract Let C be a closed, bounded, convex subset of a uniformly convex Banach space, and let $\{T_s\}$ be an asymptotic pointwise Lipschitzian semigroup of nonlinear mappings acting within C. In this chapter, we consider the implicit iteration process defined by the sequence of equations

$$x_{k+1} = c_k T_{s_{k+1}}(x_{k+1}) + (1 - c_k)x_k,$$

where each $c_k \in (0, 1)$, and the initial point $x_0 \in C$ is arbitrarily chosen. In this context, we examine the conditions under which the sequence $\{x_k\}$ converges, whether weakly or strongly, to a common fixed point of the semigroup $\{T_s\}$.

6.1 Preliminaries

As usual, let C be a nonempty, closed, bounded, and convex subset of a uniformly convex Banach space X. Another approach to fixed-point construction methods is based on the observation that for a nonexpansive mapping $T : C \to C$, for any $0 < c < 1$ and an $x_0 \in C$, the equation

$$x = cT(x) + (1 - c)x_0$$

has a unique solution $x_c \in C$, which is guaranteed by the Banach Contraction Principle and is obtained as a strong limit of the Picard iterates.

In this chapter, we extend these fixed point construction methods, often referred to as implicit iteration processes, to apply to semigroups of pointwise Lipschitzian mappings.

Let us start with the following, precise definition of the implicit iteration process and associated notions.

Definition 6.1 Let $\mathcal{T} \in \mathcal{APNS}(C)$ on C with $J = [0, +\infty)$. Assume that there exists $\gamma \geq 1$ such that $a_t(x) \leq \gamma$ for all $t \in J$ and $x \in C$. Let $0 < \beta < 1$ be chosen so that $\beta\gamma < 1$ and let $0 < \alpha < \beta$. Assume that $\{c_k\}$ is a sequence of numbers such that $0 < \alpha \leq c_k \leq \beta < 1$ for every $k \in \mathbb{N}$. Let $\{t_k\}$ is a sequence of real numbers from $(0, +\infty)$ such that

$$\sum_{n=1}^{\infty} b_{t_n}(x) < \infty$$

holds for every $x \in C$. Recall that $b_t(x) = a_t(x) - 1$. The implicit iteration process $P(C, \mathcal{T}, x_0, \{c_k\}, \{t_k\})$ is defined by the formula

$$\begin{cases} x_0 \in C \\ x_{k+1} = c_k T_{t_{k+1}}(x_{k+1}) + (1 - c_k)x_k, \ for \ k \geq 0. \end{cases} \quad (6.1)$$

We say that the sequence $\{x_k\}_{k \in \mathbb{N}_0}$ is generated by the process $P(C, \mathcal{T}, x_0, \{c_k\}, \{t_k\})$ and write

$$\{x_k\} = P(C, \mathcal{T}, x_0, \{c_k\}, \{t_k\}).$$

For $k \in \mathbb{N}_0$, $u \in C$, $w \in C$ let us introduce the notation

$$P_{k,w}(u) = c_k T_{t_{k+1}}(u) + (1 - c_k)w.$$

Observe that for every $k \in \mathbb{N}$ and all $u, v \in C$,

$$\|P_{k,w}(u) - P_{k,w}(v)\| \leq c_k \|T_{t_{k+1}}(u) - T_{t_{k+1}}(v)\| \leq c_k a_{t_{k+1}}(u)\|u - v\| \leq \beta\gamma\|u - v\|.$$

This means that each $P_{k,w} : C \to C$ is contraction. Therefore, it follows from the Banach Contraction Principle that each x_{k+1} in (6.1) is uniquely defined.

The following two results address the general behaviour of implicit iteration processes in uniformly convex Banach spaces.

Lemma 6.1 *Let X be a uniformly convex Banach space and C be a nonempty, closed, bounded, and convex subset of X. Let $\{x_k\} = P(C, \mathcal{T}, x_0, \{c_k\}, \{t_k\})$ and $w \in F(\mathcal{T})$. Then, there exists $r \geq 0$ such that $\lim_{k \to \infty} \|x_k - w\| = r$.*

Proof It follows from the inequality

$$\begin{aligned} \|x_{k+1} - w\| &= \|c_k T_{t_{k+1}}(x_{k+1}) + (1 - c_k)x_k - w\| \\ &\leq c_k \|T_{t_{k+1}}(x_{k+1}) - T_{t_{k+1}}(w)\| + (1 - c_k)\|x_k - w\| \\ &\leq c_k(1 + b_{t_k}(w))\|x_{k+1} - w\| + (1 - c_k)\|x_k - w\| \end{aligned}$$

that

$$\begin{aligned} \|x_{k+1} - w\| &\leq \frac{c_k}{1 - c_k} b_{t_k}(w)\|x_{k+1} - w\| + \|x_k - w\| \\ &\leq \frac{c_k}{1 - c_k} b_{t_k}(w)\mathrm{diam}(C) + \|x_k - w\| \\ &\leq \frac{\beta}{1 - \beta} b_{t_k}(w)\mathrm{diam}(C) + \|x_k - w\|. \end{aligned}$$

6.1 Preliminaries

It follows that for every $n \in \mathbb{N}$,

$$\|x_{k+n} - w\| \leq \|x_k - w\| + d_{k,n}, \tag{6.2}$$

where

$$d_{k,n} = \frac{\beta}{1-\beta}\mathrm{diam}(C) \sum_{i=k+1}^{k+n} b_{t_i}(w).$$

Since $\sum_{i=1}^{+\infty} b_{t_i}(w) < \infty$, we conclude that $\limsup_{k\to\infty} \limsup_{n\to\infty} d_{k,n} = 0$. Consequently, by utilising Lemma 3.1 with $r_k = \|x_k - w\|$, we infer from (6.2) that there exists an $r \in \mathbb{R}$ such that $\lim_{k\to\infty} \|x_k - w\| = r$. The proof is complete. $\square$

The following result shows that the sequence produced by the implicit iteration process in a uniformly convex Banach space represents, in a broader sense, an approximate fixed point sequence.

Lemma 6.2 *Let X be a uniformly convex Banach space and C be a nonempty, closed, bounded, and convex subset of X. Let $\{x_k\} = P(C, \mathcal{T}, x_0, \{c_k\}, \{t_k\})$. Then*

$$\lim_{k\to\infty} \|T_{t_k}(x_k) - x_k\| = 0. \tag{6.3}$$

Proof From Theorem 2.3 it follows that $F(\mathcal{T}) \neq \emptyset$. Let us fix $w \in F(\mathcal{T})$. By Lemma 6.1, there exists $r \in \mathbb{R}$ such that $\lim_{k\to\infty} \|x_k - w\| = r$. We use this to calculate

$$\limsup_{k\to\infty} \|T_{t_k}(x_k) - w\| = \limsup_{k\to\infty} \|T_{t_k}(x_k) - T_{t_k}(w)\| \tag{6.4}$$

$$\leq \limsup_{k\to\infty} a_{t_k}(w)\|x_k - w\| = r.$$

Denote $v_k = x_{k-1} - w$, $u_k = T_{t_k}(x_k) - w$ and observe that $\lim_{k\to\infty} \|v_k\| = r$. The inequality (6.4) implies that $\limsup_{k\to\infty} \|u_k\| \leq r$. Since

$$\lim_{k\to\infty} \|c_k u_k + (1 - c_k)v_k\| = \lim_{k\to\infty} \|x_k - w\| = r,$$

it follows from Lemma 3.3 that

$$\lim_{k\to\infty} \|T_{t_k}(x_k) - x_{k-1}\| = \lim_{k\to\infty} \|u_k - v_k\| = 0. \tag{6.5}$$

From (6.5) and the equality

$$\|x_{k+1} - x_k\| = \|c_k T_{t_{k+1}}(x_{k+1}) + (1 - c_k)x_k - x_k\| = c_k\|T_{t_{k+1}}(x_{k+1}) - x_k\|,$$

it follows that

$$\lim_{k\to\infty} \|x_{k+1} - x_k\| = 0. \tag{6.6}$$

By employing (6.5) and (6.6), we derive the inequality

$$\lim_{k\to\infty} \|T_{t_k}(x_k) - x_k\| \leq \lim_{k\to\infty} \|T_{t_k}(x_k) - x_{k-1}\| + \lim_{k\to\infty} \|x_{k-1} - x_k\| = 0,$$

which completes the proof. □

In the sequel, we shall require a representation of an implicit iteration process $P(C, \mathcal{T}, x_0, c_k, t_k)$ through an explicit iteration process generated by a suitably constructed asymptotic nonexpansive sequence of mappings $Z_k \in \mathcal{A}_c(C)$. The reason why the sequence Z_k must belong to $\mathcal{A}_c(C)$ will become clear in light of Lemma 6.5. As will become evident in Sect. 6.3, we need this representation primarily for the simpler, "non-pointwise" case of $\mathcal{T} \in \mathcal{ANS}(C)$. The following lemma provides the details of this representation in such a case.

Lemma 6.3 *Let C be a bounded, closed, and convex subset of a uniformly convex Banach space X. Let $\{u_k\} = P(C, \mathcal{T}, u_1, \{c_k\}, \{t_k\})$ be an implicit iteration process, where $\mathcal{T} \in \mathcal{ANS}(C)$. Then, there exists a sequence of mappings $\{Z_k\} \in \mathcal{A}_c(C)$ such that $u_{k+1} = Z_k(u_k)$ for all k, and $F(\mathcal{T}) \subset \bigcap_{k=1}^{\infty} F(Z_k)$.*

Proof For each $k \in \mathbb{N}$, define the mapping $Z_k : C \to C$ by $Z_k(w) = z_{k,w}$, where $z_{k,w}$ is the unique fixed point of the contractive mapping

$$P_{k,w}(u) = c_k T_{t_{k+1}}(u) + (1 - c_k)w.$$

By straightforward calculation we obtain $u_{k+1} = Z_k(u_k)$ for $k \in \mathbb{N}$. It is easy to prove that $F(\mathcal{T}) \subset \bigcap_{k=1}^{\infty} F(Z_k)$. We will prove that the sequence $\{Z_k\} \in \mathcal{A}_c(C)$. Let us take any $v_1, v_2 \in C$. Observe that, for ($i = 1, 2$), the equation

$$Z_k(v_i) = u_i \tag{6.7}$$

is equivalent to

$$u_i = c_k T_{t_{k+1}}(u_i) + (1 - c_k)v_i. \tag{6.8}$$

Using the notation introduced in (6.7) and (6.8), we obtain the inequality

$$\|Z_k(v_1) - Z_k(v_2)\| = \|u_1 - u_2\| \tag{6.9}$$
$$\leq c_k \|T_{t_{k+1}}(u_1) - T_{t_{k+1}}(u_2)\| + (1 - c_k)\|v_1 - v_2\|$$
$$\leq c_k a_{t_k} \|u_1 - u_2\| + (1 - c_k)\|v_1 - v_2\|.$$

It follows from (6.9) that

$$(1 - c_k a_{t_k})\|u_1 - u_2\| \leq (1 - c_k)\|v_1 - v_2\|,$$

6.1 Preliminaries

which implies that

$$\|Z_k(v_1) - Z_k(v_2)\| = \|u_1 - u_2\| \leq \frac{1-c_k}{1-c_k a_{t_k}} \|v_1 - v_2\|.$$

Since each $a_{t_k} \geq 1$ and $a_{t_k} \to 1$, it follows that each A_k, defined by

$$A_k = \frac{1-c_k}{1-c_k a_{t_k}},$$

possess the same properties. Using constants β and γ from Definition 6.1, we conclude that

$$B_k = A_k - 1 = c_k \frac{b_{t_k}}{1 - c_k a_{t_k}} \leq \frac{\beta}{1-\beta\gamma} b_{t_k}. \tag{6.10}$$

Since $\beta\gamma < 1$ and $\sum_{k=1}^{\infty} b_{t_k} < \infty$, it follows from (6.10) that $\sum_{k=1}^{\infty} B_k < \infty$. Note that this implies that $\prod_{k=1}^{\infty} A_k < \infty$. Finally, we conclude that the sequence $\{Z_k\}$ belongs to $\mathcal{A}_c(C)$, as claimed. □

Remark 6.1 An attentive reader may have noticed why we assumed $\mathcal{T} \in \mathcal{ANS}(C)$ instead of the more general $\mathcal{T} \in \mathcal{APNS}(C)$. If a_{t_k} were not constants, then in inequality (6.9), we would have $a_{t_k}(u_k)$ instead of $a_{t_k}(v_k)$, introducing an implicit effect that we aim to avoid in this explicit representation.

Remark 6.2 Note that the definition of each mappings Z_k depends only on c_k and $T_{t_{k+1}}$. Specifically, it does not depend on the starting point u_1 of the implicit iteration process $P(C, \mathcal{T}, u_1, \{c_k\}, \{t_k\})$.

Observe that the properties of the implicit iteration processes, discussed so far, did not depend on our choice of the sequence $\{t_k\}$. Not surprisingly, in order to be able to prove the convergence of such processes we will have to impose some restrictions on $\{t_k\}$. This leads us to a notion of a normalised implicit iteration process.

Definition 6.2 We will say that an implicit iteration process

$$\{x_k\} = P(C, \mathcal{T}, x_0, \{c_k\}, \{t_k\})$$

is normalised if the following two conditions are satisfied,

$$\lim_{k \to \infty} t_k = 0 \tag{6.11}$$

and

$$\lim_{k \to \infty} \frac{1}{t_k} \|T_{t_k}(x_k) - x_k\| = 0. \tag{6.12}$$

Proposition 6.1 *It is worthwhile noting that we can always construct a sequence $\{t_k\}$ satisfying (6.11) and (6.12). Let us sketch one particular method of such a construction which has been studied and used in relation to the convergence of processes in Hilbert spaces and in uniformly Banach spaces with the Opial property. Let us use a starting sequence of positive real numbers $\{s_n\}$ such that*

$$\liminf_{n\to\infty} s_n = 0, \ \limsup_{n\to\infty} s_n > 0$$

and

$$\lim_{n\to\infty}(s_{n+1} - s_n) = 0.$$

Proof Fix $0 < \eta < \limsup_{n\to\infty} s_n$. Assume that x_{k-1} have been constructed. Let us construct now t_k. In view of (6.3) and the fact that $\frac{\eta}{2^k}$ is a cluster point for the sequence $\{s_n\}$ (by Lemma 3.2), it follows that there exists $j \in \mathbb{N}$ such that

$$\frac{\eta}{2^{k+1}} < s_j < \frac{\eta}{2^{k-1}} \tag{6.13}$$

and

$$\|T_{s_j}(y_j) - y_j\| < \frac{\eta}{2^{2(k+1)}}, \tag{6.14}$$

where y_j is constructed using c_{k-1} and x_{k-1}. Put $t_k = s_j$ and $x_k = y_j$. Use (6.14) and (6.13) to verify (6.11) and (6.12). □

The role of sequences defining the normalised process becomes clear in view of the following corollary to the Demiclosedness Principles proved in Chap. 3.

Lemma 6.4 *Let C be a nonempty, closed, bounded, and convex subset of a uniformly convex Banach space X. Assume that either X has the Opial property or that X is uniformly smooth. Let $\mathcal{T} \in \mathcal{APNS}(C)$ be equicontinuous on C. Assume that there exists $\gamma \geq 1$ such that $a_t(x) \leq \gamma$ holds for every $x \in C$ and for all $t \in J = [0, +\infty)$. Let $\{s_i\}$ denote a sequence of strictly positive real numbers converging to zero as $i \to \infty$. Additionally, let $\{x_i\}$ be a sequence of elements from C such that $\{x_i\}$ converges weakly to some element $w \in C$. If*

$$\lim_{i\to\infty} \frac{1}{s_i}\|T_{s_i}(x_i) - x_i\| = 0, \tag{6.15}$$

then $w \in F(\mathcal{T})$.

Proof Note that, according to Theorem 2.3, $F(\mathcal{T})$ is not empty. Observe also that the sequence $\{x_i\}$ is regular with respect to $\mathcal{T}$ because of the boundedness of each function a_t. Let us fix an arbitrary $t > 0$. Without any loss of generality we can assume that $s_i < t$ for every $i \in \mathbb{N}$. Denoting for simplicity $p_i = \left\lfloor \frac{t}{s_i} \right\rfloor$ we have

$$\|x_i - T_t(x_i)\| \leq \sum_{k=0}^{p_i-1} \|T_{(k+1)s_i}(x_i) - T_{ks_i}(x_i)\| + \|T_{p_i s_i}(x_i) - T_t(x_i)\|$$

$$\leq \|T_{s_i}(x_i) - x_i\| \sum_{k=0}^{p_i-1} a_{ks_i}(x_i) + \|T_{p_i s_i}(x_i) - T_t(x_i)\|$$

$$\leq \frac{t}{s_i}\gamma\|T_{s_i}(x_i) - x_i\| + \gamma\|T_{t-p_i s_i}(x_i) - x_i\|.$$

Since $\frac{t}{s_i}\|T_{s_i}(x_i) - x_i\| \to 0$ by (6.15) and $t - p_i s_i \to 0$ it follows from the equicontinuity of $\mathcal{T}$ that

$$\|x_i - T_t(x_i)\| \to 0.$$

Therefore, the sequence $\{x_i\}$ is an approximate fixed point sequence for every $t \in J$. By assumption, $x_i \rightharpoonup w$. Consequently, it follows from the Demiclosedness Principle 1 (Theorem 3.4) in the context of the Opial property case, or from Demiclosedness Principle 2 (Theorem 3.5) in the uniform smoothness case, that $T_t(w) = w$ for all $t \in J$, as claimed. $\square$

6.2 Weak Convergence of Implicit Iteration Processes in Uniformly Convex Banach Spaces with the Opial Property

At this point, we have gathered all the necessary preparatory information to prove our weak convergence theorem for normalised implicit iteration processes in uniformly convex Banach spaces that possess the Opial property.

Theorem 6.1 *Let C be a nonempty, closed, bounded, and convex subset of a uniformly convex Banach space X with the Opial property. Let $\mathcal{T} \in \mathcal{APNS}(C)$ be equicontinuous on C. Assume that there exists $\gamma \geq 1$ such that $a_t(x) \leq \gamma$ holds for every $x \in C$ and for all $t \in J = [0, +\infty)$. Let $\{x_k\} = P(C, \mathcal{T}, x_0, \{c_k\}, \{t_k\})$ be a normalised implicit iteration process. Then, there exists a common fixed point $w \in F(\mathcal{T})$ such that $x_k \rightharpoonup w$.*

Proof Since $P(C, \mathcal{T}, x_0, \{c_k\}, \{t_k\})$ is normalised it follows that $t_k \to 0$ and that

$$\lim_{k\to\infty} \frac{1}{t_k}\|T_{t_k}(x_k) - x_k\| = 0, \tag{6.16}$$

which implies that

$$\lim_{k\to\infty} \|T_{t_k}(x_k) - x_k\| = 0.$$

Consider $y, z \in C$, which are two weak cluster points of the sequence $\{x_k\}$. Then, there exist two subsequences $\{y_k\}$ and $\{z_k\}$ of $\{x_k\}$ such that $y_k \rightharpoonup y$ and $z_k \rightharpoonup z$. It follows from Lemma 6.4 that $y, z \in F(\mathcal{T})$. According to Lemma 6.1, the following limits exist:

$$r_1 = \lim_{k\to\infty} \|x_k - y\|, \ r_2 = \lim_{k\to\infty} \|x_k - z\|.$$

We assert that $y = z$. For contradiction, assume $y \neq z$. By the Opial property,

$$\begin{aligned} r_1 &= \liminf_{k\to\infty} \|y_k - y\| < \liminf_{k\to\infty} \|y_k - z\| = r_2 \\ &= \liminf_{k\to\infty} \|z_k - z\| < \liminf_{k\to\infty} \|z_k - y\| = r_1. \end{aligned}$$

This contradiction indicates that $y = z$, showing that the sequence $\{x_k\}$ has no more than one weak cluster point. Since C is weakly sequentially compact, it follows that $\{x_k\}$ indeed has exactly one weak cluster point $w \in C$, meaning that that $x_k \rightharpoonup w$. Since (6.16) holds, we can apply Lemma 6.4 to conclude that $w \in \mathcal{T}$. This completes the proof. □

6.3 Weak Convergence of Implicit Iteration Processes in Uniformly Convex and Uniformly Smooth Banach Spaces

The scenario becomes more intricate when the assumption of the Opial property is substituted with the uniform smoothness of X. To start, we require the following result, which is analogous to Lemmas 4.4 and 5.4 that address the generalised Krasnosel'skii-Mann and Ishikawa processes, respectively. We will apply Lemma 6.3, which explains why in this whole section we assume that $\mathcal{T} \in \mathcal{ANS}(C)$.

Lemma 6.5 *Let C be a bounded, closed, and convex subset of a uniformly convex and uniformly smooth Banach space X. Assume that $\{x_k\} = P(C, \mathcal{T}, x_0, \{c_k\}, \{t_k\})$ is a normalised implicit iteration process, where $\mathcal{T} \in \mathcal{ANS}(C))$ is equicontinuous on C. If $w_1, w_2 \in F(\mathcal{T})$ then*

$$\langle y - z, J(w_1 - w_2)\rangle = 0 \tag{6.17}$$

holds for any two weak cluster points y, z of the sequence $\{x_k\}$.

Proof Let $Z_k : C \to C$ be the mappings defined in Lemma 6.3. Recall that we established there that $x_{k+1} = Z_k(x_k)$, $F(\mathcal{T}) \subset \bigcap_{k=1}^{\infty} F(Z_k)$, and that the sequence $\{Z_k\}$ belongs to $\mathcal{A}_c(C)$. Therefore, by applying Lemma 3.8, we conclude that the equality (6.17) holds for any two weak cluster points y, z of the sequence $\{x_k\}$. □

We are ready to demonstrate the main weak convergence result of this section.

Theorem 6.2 *Let C be a bounded, closed, and convex subset of a uniformly convex and uniformly smooth Banach space X. Let $\mathcal{T}$ be an asymptotic nonexpansive semigroup ($\mathcal{T} \in \mathcal{ANS}(C)$), which is equicontinuous on C. Assume that there exists $\gamma \geq 1$ such that $a_t(x) \leq \gamma$ holds for every $x \in C$ and for all $t \in J = [0, +\infty)$. If*

$\{x_k\} = P(C, \mathcal{T}, x_0, \{c_k\}, \{t_k\})$ is a normalised implicit iteration process, then there exists a common fixed point $w \in F(\mathcal{T})$ such that $x_k \rightharpoonup w$.

Proof Since $P(C, \mathcal{T}, x_0, \{c_k\}, \{t_k\})$ is normalised it follows that $t_k \to 0$ and that

$$\lim_{k \to \infty} \frac{1}{t_k} \|T_{t_k}(x_k) - x_k\| = 0. \tag{6.18}$$

Consider $y, z \in C$ two weak cluster points of the sequence $\{x_k\}$. There exist two subsequences $\{x_{\alpha_n}\}$ and $\{x_{\beta_n}\}$ of the sequence $\{x_k\}$ such that $x_{\alpha_n} \rightharpoonup y$, $x_{\beta_n} \rightharpoonup z$. From (6.18) we conclude that

$$\lim_{n \to \infty} \frac{1}{t_{\alpha_n}} \|T_{t_{\alpha_n}}(x_{\alpha_n}) - x_{\alpha_n}\| = 0 \text{ and } \lim_{n \to \infty} \frac{1}{t_{\beta_n}} \|T_{t_{\beta_n}}(x_{\beta_n}) - x_{\beta_n}\| = 0.$$

By Lemma 6.4, we have $y \in F(\mathcal{T})$ and $z \in F(\mathcal{T})$. According to Lemma 6.5, this implies $\|y - z\|^2 = \langle y - z, J(y - z) \rangle = 0$, leading us to conclude that $y = z$. Therefore, the sequence $\{x_k\}$ has at most one weak cluster point. Since C is weakly sequentially compact, it follows that $\{x_k\}$ has exactly one weak cluster point $w \in C$. This means that $x_k \rightharpoonup w$. It follows from Lemma 6.4 that $w \in F(\mathcal{T})$. The proof is now complete. □

6.4 Strong Convergence of Implicit Iteration Processes

The proceedings become much more straightforward when the set C is compact in the norm topology.

Theorem 6.3 *Let C be a convex compact subset of a uniformly convex and uniformly smooth Banach space X. Let $\mathcal{T}$ be an asymptotic pointwise nonexpansive semigroup. Assume that there exists $\gamma \geq 1$ such that $a_t(x) \leq \gamma$ holds for every $x \in C$ and for all $t \in J = [0, +\infty)$. Let us assume that $\{x_k\} = P(C, \mathcal{T}, x_0, \{c_k\}, \{t_k\})$ is an implicit iteration process. Assume that the sequence $\{t_k\}$ satisfies conditions*

$$0 = \liminf_{n \to \infty} t_n < \limsup_{n \to \infty} t_n,$$

$$\lim_{n \to \infty} (t_{n+1} - t_n) = 0.$$

Then, there exists a common fixed point $x \in F(\mathcal{T})$ such that $\|x_k - x\| \to 0$.

Proof Fix any $0 < t < \limsup_{n \to \infty} t_n$. According to Lemma 3.2, there exists $\{t_{k_n}\}$ a subsequence of $\{t_k\}$ such that

$$\lim_{n \to \infty} t_{k_n} = t. \tag{6.19}$$

From Lemma 6.2 it follows that

$$\lim_{n \to \infty} \|T_{t_{k_n}}(x_{k_n}) - x_{k_n}\| = 0. \tag{6.20}$$

Since C is compact, there exists $\{x_{k_{n_i}}\}$, as subsequence of $\{x_{k_n}\}$ and an element of $x \in C$ such that

$$\lim_{i \to \infty} \|T_{t_{k_{n_i}}}(x_{k_{n_i}}) - x\| = 0. \tag{6.21}$$

Denote $s_i = t_{k_{n_i}}$, $w_i = x_{k_{n_i}}$. Observe that from (6.20), (6.21) and the inequality

$$\|w_i - x\| \le \|w_i - T_{s_i}(w_i)\| + \|T_{s_i}(w_i) - x\|,$$

we can derive $\|w_i - x\| \to 0$. Therefore,

$$\begin{aligned}\|T_{s_i}(x) - x\| &\le \|T_{s_i}(x) - T_{s_i}(w_i)\| + \|T_{s_i}(w_i) - w_i\| + \|w_i - x\| \quad (6.22)\\ &\le a_{s_i}(x)\|x - w_i\| + \|T_{s_i}(w_i) - w_i\| + \|w_i - x\|\\ &\le \gamma \|x - w_i\| + \|T_{s_i}(w_i) - w_i\| + \|w_i - x\| \to 0,\end{aligned}$$

as $i \to \infty$. From (6.22), (6.19), and the continuity of the semigroup $\mathcal{T}$, we conclude that

$$\|T_t(x) - x\| \le \|T_t(x) - T_{s_i}(x)\| + \|T_{s_i}(x) - x\| \to 0,$$

as $i \to \infty$. Thus, $T_t(x) = x$. We need now to prove this equality for any $s > 0$. Observe that there exist $0 < t < \limsup_{n \to \infty} t_n$, $0 < u < \limsup_{n \to \infty} t_n$, and $k \in \mathbb{N}_0$ such that $s = t + ku$. Therefore, $T_s(x) = x$, because

$$T_s(x) = T_{ku}(T_t(x)) = T_{ku}(x) = T_{u + \cdots + u}(x) = x.$$

We conclude that $x \in F(\mathcal{T})$. It remains to show that $\lim_{k \to \infty} \|x_k - x\| = 0$. To achieve this, note that, according to Lemma 6.1, there exists a non-negative number r such that $\lim_{k \to \infty} \|x_k - x\| = r$. We previously established that

$$\lim_{i \to \infty} \|x_{k_{n_i}} - x\| = \lim_{i \to \infty} \|w_i - x\| = 0,$$

which necessitates that $r = 0$. Thus, the proof is complete. □

6.5 Notes

There are known results on weak or strong convergence of implicit iteration processes for nonexpansive semigroups in Hilbert spaces and uniformly convex Banach spaces with the Opial property, refer to works by Suzuki [99], Xu [108], Kim and Takahashi [48], Saejung [96], Thong [104]. The question of weak convergence when X is simultaneously uniformly convex and uniformly smooth was resolved by Kozlowski

6.5 Notes

in [67]. The strong convergence under a compactness assumption of C in a uniformly convex Banach spaces was examined by Kozlowski in [64]. Since the seminal paper by Goebel and Kirk [40], the concept of asymptotic nonexpansive mappings and its extension to semigroups have been firmly established in mainstream fixed point theory and its applications. The stability of implicit iteration processes in this case is discussed in Kozlowski's recent work [76]. The convergence results for the implicit iteration processes can be extended beyond the scope of Banach spaces. For example, the recently published paper [73] considers the case of semigroups of operators acting within modular spaces.

Chapter 7
Stability of Common Fixed Point Construction Processes

Abstract Consider a sequence of pointwise Lipchitzian operators $\{H_k\}$, and define a generalised iteration process as follows: starting from an arbitrary point $u_1 \in C$, generate a sequence $\{u_k\}$ by iteratively computing $u_{k+1} = H_k(u_k)$ for $k \geq 1$. If this process converges weakly to a common fixed point of a semigroup $\mathcal{T}$, regardless of the starting point, it serves as a method for approximating such common fixed points. In practical implementations, each iteration may introduce computational errors. Therefore, it is crucial to assess the stability of this process—specifically, whether it continues to converge weakly to a common fixed point when each iteration k produces a point x_{k+1} that is close to, but not exactly equal to $H_k(x_k)$. This chapter focuses on stability under summable errors, meaning that for any sequence $\{x_k\}$ within C where $\sum_{k=1}^{\infty} \|x_{k+1} - H_k(x_k)\| < \infty$, the sequence converges in the weak topology to a common fixed point of $\mathcal{T}$, assuming that the sequence $u_{k+1} = H_k(u_k)$ is weak-convergent to a (possibly different) common fixed point of $\mathcal{T}$ for any initial element $u_1 \in C$. This chapter aims to present a general method for establishing a specific type of stability in iteration processes that converge to common fixed points of pointwise Lipschitzian semigroups. We illustrate this method using examples from the weak convergence results of generalised Krasnosel'skii-Mann, Ishikawa, and implicit iteration processes discussed in the previous chapters. These examples serve as patterns that can be adapted for various other contexts.

7.1 Stability of Asymptotic Pointwise Nonexpansive Sequences

In this section, we assume that the sequences $\{H_k\} \in \mathcal{A}(C)$ are as general as possible. However, it is important to note that these sequences are intended to reflect well-known fixed-point iteration algorithms, including the generalised Krasnosel'skii-Mann, generalised Ishikawa, and implicit iteration processes, which will be explored in detail in the subsequent sections.

Let $\{H_k\}$ be an asymptotic pointwise nonexpansive sequence belonging to $\mathcal{A}(C)$ (recall Definition 3.6). We will use the following notation:

$S_k^n = H_n \circ H_{n-1} \circ \ldots \circ H_k$, where $n \geq k$ are natural numbers. Note that $S_k^k = H_k$. By convention we set $S_k^{k-1} = Id$, where Id stands for the identity operator.

Let us start with the following two technical results.

Lemma 7.1 *Assume that* $\{H_k\} \in \mathcal{A}(C)$. *Let* $\{x_k\}$ *be a sequence of elements from* C. *Then, the inequality*

$$\|S_{k+q+1}^{k+q+i}(x_{k+q+1}) - S_k^{k+q+i}(x_k)\| \tag{7.1}$$
$$\leq \prod_{j=1}^{i} A_{k+q+j}(S_{k+q+1}^{k+q+j-1}(x_{k+q+1})) \|x_{k+q+1} - S_k^{k+q}(x_k)\|$$

holds for any three natural numbers $k, q, i \in \mathbb{N}$.

Proof Let us fix $k, q \in \mathbb{N}$ and proceed by induction with respect to $i \in \mathbb{N}$. For $i = 1$, we have

$$\|S_{k+q+1}^{k+q+1}(x_{k+q+1}) - S_k^{k+q+1}(x_k)\| = \|H_{k+q+1}(x_{k+q+1}) - H_{k+q+1}(S_k^{k+q}(x_k))\|$$
$$\leq A_{k+q+1}(x_{k+q+1}) \|x_{k+q+1} - S_k^{k+q}(x_k)\|$$
$$= A_{k+q+1}(S_{k+q+1}^{k+q}(x_{k+q+1})) \|x_{k+q+1} - S_k^{k+q}(x_k)\|.$$

Assume now that (7.1) is true for a natural number $i - 1$, $i \geq 2$. Let us show it also holds for i. We have the following sequence of inequalities,

$$\|S_{k+q+1}^{k+q+i}(x_{k+q+1}) - S_k^{k+q+i}(x_k)\|$$
$$= \|H_{k+q+i}(S_{k+q+1}^{k+q+i-1}(x_{k+q+1})) - H_{k+q+i}(S_k^{k+q+i-1}(x_k))\|$$
$$\leq A_{k+q+i}(S_{k+q+1}^{k+q+i-1}(x_{k+q+1})) \|S_{k+q+1}^{k+q+i-1}(x_{k+q+1}) - S_k^{k+q+i-1}(x_k)\|$$
$$\leq A_{k+q+i}(S_{k+q+1}^{k+q+i-1}(x_{k+q+1})) \prod_{j=1}^{i-1} A_{k+q+j}(S_{k+q+1}^{k+q+j-1}(x_{k+q+1})) \|x_{k+q+1} - S_k^{k+q}(x_k)\|$$
$$= \prod_{j=1}^{i} A_{k+q+j}(S_{k+q+1}^{k+q+j-1}(x_{k+q+1})) \|x_{k+q+1} - S_k^{k+q}(x_k)\|,$$

proving the inductive step. □

Lemma 7.2 *Assume that* $\{H_k\} \in \mathcal{A}(C)$. *Let* $\{x_k\}$ *be a sequence of elements from* C *and* $\{r_k\}$ *be a sequence of positive numbers such that for every* $k \in \mathbb{N}$,

$$\|x_{k+1} - H_k(x_k)\| \leq r_k. \tag{7.2}$$

Then,

$$\|x_{k+i+1} - S_k^{k+i}(x_k)\| \leq \left(\prod_{j=k}^{k+i} A_j(x_j)\right)\left(\sum_{j=k}^{k+i} r_j\right) \tag{7.3}$$

holds for every $k \in \mathbb{N}$ *and every* $i \in \mathbb{N}_0$.

7.1 Stability of Asymptotic Pointwise Nonexpansive Sequences

Proof Let us fix $k \in \mathbb{N}$ and proceed by induction with respect to $i \in \mathbb{N}_0$. Let $i = 0$ and observe that by (7.2) and the fact that each $A_k(x_k) \geq 1$,

$$\|x_{k+1} - S_k^k(x_k)\| = \|x_{k+1} - H_k(x_k)\| \leq r_k \leq A_k(x_k)r_k,$$

which shows that (7.3) holds for $i = 0$. Let us assume now that (7.3) is true for $i - 1$, where $i \in \mathbb{N}$ and proceed with the proof of the inductive step. Calculate

$$\begin{aligned}\|x_{k+i+1} - S_k^{k+i}(x_k)\| &\leq \|x_{k+i+1} - H_{k+i}(x_{k+i})\| + \|H_{k+i}(x_{k+i}) - S_k^{k+i}(x_k)\| \\ &\leq r_{k+i} + \|H_{k+i}(x_{k+i}) - H_{k+i}(S_k^{k+i-1}(x_k))\| \\ &\leq r_{k+i} + A_{k+i}(x_{k+i})\|x_{k+i} - S_k^{k+i-1}(x_k)\| \\ &\leq r_{k+i} + A_{k+i}(x_{k+i})\left(\prod_{j=k}^{k+i-1} A_j(x_j)\right)\left(\sum_{j=k}^{k+i-1} r_j\right) \\ &\leq \left(\prod_{j=k}^{k+i} A_j(x_j)\right)r_{k+i} + \left(\prod_{j=k}^{k+i} A_j(x_j)\right)\left(\sum_{j=k}^{k+i-1} r_j\right) \\ &= \left(\prod_{j=k}^{k+i} A_j(x_j)\right)\left(\sum_{j=k}^{k+i} r_j\right),\end{aligned}$$

where we used (7.2), the inductive hypothesis, and that $\prod_{j=k}^{k+i} A_j(x_j) \geq 1$ for any $k \in \mathbb{N}$ and $i \in \mathbb{N}_0$. The proof is complete. $\square$

Before employing the two technical lemmas in the proof of our general stability results, Theorems 7.1 and 7.2, we introduce the following controllability concepts for a sequence of pointwise Lipschitzian operators $\{H_k\} \in \mathcal{A}(C)$.

Definition 7.1 Let $\{x_k\}$ be a sequence of elements from C. Let $\{H_k\} \in \mathcal{A}(C)$. Assume that Σ is a subset of C. The sequence $\{H_k\}$ will be called well-controlled over Σ if there exists a natural number p_0 and a constant $1 \leq L < +\infty$ such that

$$\prod_{p=p_0}^{\infty} L_p \leq L < +\infty, \tag{7.4}$$

where

$$L_p = \sup\{A_p(z) : z \in \Sigma\}. \tag{7.5}$$

Let us introduce the following notation.

Definition 7.2 Given $\{H_k\} \in \mathcal{A}(C)$ and $x \in C$, denote

$$\text{Iter}(\{H_k\}, x) = \{S_m^{m+j-2}(x) : m, j \in \mathbb{N}\}.$$

Note that $Iter(\{H_k\}, x)$ is simply a subset of C consisting of all iterations of the form $H_{m+j-2} \circ H_{m+j-1} \circ ... \circ H_m(x)$.

Definition 7.3 The sequence $\{H_k\} \in \mathcal{A}(C)$ will be called well-controlled if it is well-controlled over every $\Sigma \subset C$ such that $\bigcap_{x \in C}^{\infty} \text{Iter}(\{H_k\}, x) \subset \Sigma$.

Remark 7.1 The technical assumptions on $\{H_k\} \in \mathcal{A}(C)$ for being well-controlled as outlined in formulas (7.4) and (7.5) may, at the first sight, appear non-intuitive. However, they are important for two main reasons. Firstly, we aim to minimise restrictions on the functions $A_p(\cdot)$ as much as possible. Secondly, we require flexibility to address a diverse range of scenarios. In the following sections, we will demonstrate that when these assumptions are applied to specific iterative processes, they often translate into more natural and intuitive conditions.

Theorem 7.1 Let $\mathcal{T} \in \mathcal{APNS}(C)$. Let $\{x_k\}$ be a sequence of elements from C and let $\{H_k\} \in \mathcal{A}(C)$. Let $\Sigma \subset C$ be such that $\bigcap_{k=k_0}^{\infty} Iter(\{H_n\}, x_k) \subset \Sigma$ for some $k_0 \in \mathbb{N}$. Assume that $\{H_k\}$ is well-controlled over Σ. In addition, assume that for every $k \in \mathbb{N}$ there exists a common fixed point $y_k \in F(\mathcal{T})$ such that

$$S_k^{k+n-2}(x_k) \rightharpoonup y_k, \tag{7.6}$$

as $n \to \infty$. If

$$\sum_{j=1}^{\infty} \|x_{k+1} - H_k(x_k)\| < +\infty, \tag{7.7}$$

then there exists $w \in F(\mathcal{T})$ such that $x_k \rightharpoonup w$ as $k \to +\infty$.

Proof Let p_0, L_p and L are selected as per Definition 7.1. Let $k \geq \max\{k_0, p_0\}$. Take any two natural numbers $q, i \in \mathbb{N}$. It follows from the assumption that $\{H_k\}$ is well-controlled over Σ that

$$\prod_{j=1}^{i} A_{k+q+j}(S_{k+q+1}^{k+q+j-1}(x_{k+q+1})) \leq \prod_{p=p_0}^{\infty} L_p \leq L < +\infty, \tag{7.8}$$

and

$$\prod_{j=k}^{k+i} A_j(x_j) \leq \prod_{p=p_0}^{\infty} L_p \leq L < +\infty. \tag{7.9}$$

Denote $r_k = \|x_{k+1} - H_k(x_k)\|$. Using (7.8) and (7.9) along with Lemma 7.1 and Lemma 7.2, we obtain the following estimation:

$$\|S_{k+q+1}^{k+q+i}(x_{k+q+1}) - S_k^{k+q+i}(x_k)\| \leq L\|x_{k+q+1} - S_k^{k+q}(x_k)\| \leq L^2 \sum_{j=k}^{k+q} r_j \leq \delta_k, \tag{7.10}$$

7.1 Stability of Asymptotic Pointwise Nonexpansive Sequences

where $\delta_k = L^2 \sum_{j=k}^{\infty} r_j$. Note that $\delta_k \to 0$ as $k \to +\infty$ because $\sum_{j=1}^{\infty} r_j < +\infty$ by (7.7). Recall that Mazur's Theorem implies that if $z_n \rightharpoonup z$ and all $z_n \in B(0, \delta)$ then $z \in B(0, \delta)$. Since, by (7.6),

$$S_{k+q+1}^{k+q+i}(x_{k+q+1}) \rightharpoonup y_{k+q+1}$$

and

$$S_k^{k+q+i}(x_k) \rightharpoonup y_k$$

as $i \to +\infty$, it follows from (7.10), by Mazur's Theorem, that $\|y_{k+q+1} - y_k\| \leq \delta_k$ holds for every $k \in \mathbb{N}$. Therefore,

$$\|y_{k+q+1} - y_k\| \leq \delta_k \to 0, \tag{7.11}$$

as $k \to \infty$, and we can conclude that $\{y_k\}$ is a Cauchy sequence. From the completeness of X, it follows that there exists $w \in C$ such that $w = \lim_{k \to +\infty} y_k$. Moreover, by taking q to infinity in (7.11) we get

$$\|y_k - w\| \leq \delta_k \to 0.$$

Let us prove now that w is indeed a common fixed point for $\mathcal{T}$. Let us fix an arbitrary $t \in J$. Remembering that $\mathcal{T}$ is a pointwise Lipschitzian semigroup and that y_k is a common fixed point for every $k \in \mathbb{N}$, we can see that

$$\begin{aligned} \|T_t(w) - w\| &\leq \|T_t(w) - T_t(y_k)\| + \|T_t(y_k) - y_k\| + \|y_k - w\| \\ &\leq a_t(w)\|w - y_k\| + \|y_k - w\| \\ &\leq (a_t(w) + 1)\|y_k - w\| \\ &\leq (a_t(w) + 1)\delta_k \to 0, \end{aligned}$$

whenever $k \to +\infty$.

To finish the proof we still need to show that $x_k \rightharpoonup w$ as $k \to +\infty$. Let φ be a continuous linear functional on X with $\|\varphi\|_{X^*} \leq 1$ and let $\varepsilon > 0$ be arbitrarily fixed. Let us fix temporarily $k, i \in \mathbb{N}$. Let $k_\varepsilon \geq p_0$ be such that $\delta_k \leq \frac{\varepsilon}{3}$ for $k \geq k_\varepsilon$. Observe that, in view of (7.10), $\|S_k^{k+i}(x_k) - x_{k+i+1}\| \leq \frac{\delta_k}{L} \leq \delta_k$. Therefore, for $k \geq k_\varepsilon$, we have

$$\begin{aligned} |\varphi(w - x_{k+i+1})| &\leq |\varphi(w - y_k)| + |\varphi(y_k - S_k^{k+i}(x_k))| \\ &\quad + |\varphi(S_k^{k+i}(x_k) - x_{k+i+1})| \\ &\leq \|w - y_k\| + |\varphi(y_k - S_k^{k+i}(x_k))| + \delta_k \\ &\leq \frac{\varepsilon}{3} + |\varphi(y_k - S_k^{k+i}(x_k))| + \frac{\varepsilon}{3}. \end{aligned} \tag{7.12}$$

Since $S_k^{k+i}(x_k) \rightharpoonup y_k$ as $i \to +\infty$, there exists $i_0 \in \mathbb{N}$ such that for any natural $i \geq i_0$,

$$|\varphi(y_k - S_k^{k+i}(x_k))| \leq \frac{\varepsilon}{3}.$$

This, together with (7.12), proves that $x_k \rightharpoonup w$. The proof is complete. □

Consider a sequence $\{H_k\} \in \mathcal{A}(C)$. We introduce a generalised iterative procedure derived from this sequence, defined as follows.

Definition 7.4 Let $\{H_k\} \in \mathcal{A}(C)$. An iterative process generated by $\{H_k\}$, denoted by $\Gamma(\{H_k\})$, is defined as the process producing a sequence $\{u_k\}$ by setting an arbitrary starting point $u_1 \in C$ and calculating $u_{k+1} = H_k(u_k)$ for $k \geq 1$.

Definition 7.5 Let $\mathcal{T} \in \mathcal{APNS}(C)$ and $\{H_k\} \in \mathcal{A}(C)$. We say that an iterative process $\Gamma(\{H_k\})$ is weak-convergent to $F(\mathcal{T})$ if for every $x \in C$ and every $k \in \mathbb{N}$, the sequence $\{S_k^{k+n-2}(x)\}_{n \in \mathbb{N}}$ converges in the weak topology to an element of $F(\mathcal{T})$.

Remark 7.2 Observe that the weak-convergence to $F(\mathcal{T})$ of an iterative process $\Gamma(\{H_k\})$ means simply that whatever the starting point $x \in C$ is chosen, the subsequent iterations of $H_k(x)$ will weakly converge to some common fixed point of $\mathcal{T}$. Recall that our weak convergence theorems for the generalised Krasnosel'skii-Mann, generalised Ishikawa, and the implicit iterative processes articulate precisely this phenomenon. However, it is important to recognise that, generally, pointwise Lipschitzian semigroups do not guarantee the uniqueness of common fixed points. Consequently, the result of the construction of a common fixed point, as outlined in Definition 7.5, may depend on the selection of $x \in C$ and $k \in \mathbb{N}$.

Having an iterative process $\Gamma(\{H_k\})$, which is weak-convergent to $F(\mathcal{T})$, provides a method for approximating common fixed points of the semigroup $\mathcal{T}$. However, in algorithmic implementations, the computations at each iteration step typically introduce errors. Therefore, it is essential to assess whether the process $\Gamma(\{H_k\})$ remains stable, meaning whether it continues to weakly converge to a common fixed point when, at each iteration step k, we select x_{k+1} to be close to but not exactly equal to $H_k(x_k)$. To be more precise, let us introduce the following definition.

Definition 7.6 We define a generalised iteration process $\Gamma(\{H_k\})$, which weakly converges to $F(\mathcal{T})$, as being stable under summable errors if for every sequence $\{x_k\}$ satisfying the estimation

$$\sum_{k=1}^{\infty} \|x_{k+1} - H_k(x_k)\| < \infty,$$

there exists an $w \in F(\mathcal{T})$ such that $x_k \rightharpoonup w$.

Using the notion of stability and the notation introduced above, we can derived from Theorem 7.1 the following general result.

Theorem 7.2 Let $\mathcal{T} \in \mathcal{APNS}(C)$ and let $\{H_k\} \in \mathcal{A}(C)$ be well-controlled. Assume that the iterative process $\Gamma(\{H_k\})$ is weak-convergent to $F(\mathcal{T})$. Then $\Gamma(\{H_k\})$ is stable under summable errors.

The framework developed in this section offers a versatile foundation that facilitates the application of the general stability results, Theorems 7.1 and 7.2, to various scenarios where the existence of common fixed points and convergence of the iteration process in the weak topology are assured. In the subsequent sections, we will utilise these general results to address the questions of stability in the presence of summable errors, specifically in the context of the iteration processes introduced in the earlier chapters.

7.2 Stability of Krasnosel'skii-Mann Processes

In this section, we will demonstrate the stability of the generalised Krasnosel'skii-Mann iteration processes using the general results obtained in the previous section. We assume that the set $\Sigma \subset C$ is such that $\bigcap_{k=k_0}^{\infty} \text{Iter}(\{H_n\}, x_k) \subset \Sigma$ for some $k_0 \in \mathbb{N}$. We also assume that $\{H_k\}$ is well-controlled over Σ.

Lemma 7.3 *Let C be a bounded, closed, and convex subset of a uniformly convex Banach space X, and let $\mathcal{T} \in \mathcal{APNS}(C)$. Consider a well-defined generalised Krasnosel'skii-Mann process $gKM(\mathcal{T}, \{c_k\}, \{t_k\})$. Let for every $k \in \mathbb{N}$, H_k be defined by*

$$H_k(x) = c_k T_{t_k}(x) + (1 - c_k)x.$$

Denote

$$B_{t_n} = \sup\{b_{t_n}(z) : z \in \Sigma\}.$$

Assume that there exists $n_0 \in \mathcal{N}$ such that

$$B := \sum_{n=n_0}^{\infty} B_{t_n} < \infty. \tag{7.13}$$

Then, $\{H_k\}$ is a sequence of pointwise Lipschitzian operators, which is well-controlled over Σ.

Proof Observe that for every $k \in \mathbb{N}$ and all $x, y \in C$,

$$\|H_k(x) - H_k(y)\| \le c_k \|T_{t_k}(x) - T_{t_k}(y)\| + (1 - c_k)\|x - y\|$$
$$\le (c_k a_{t_k}(x) + 1 - c_k)\|x - y\|.$$

We conclude that the sequence $\{H_k\}$ is a sequence of pointwise Lipschitzian operators with $A_k(x) = 1 + c_k b_{t_k}(x) \ge 1$. Observe that in this situation,

$$L_n = \sup\{A_n(z) : z \in \Sigma\} \le 1 + \sup\{b_{t_n}(x) : z \in \Sigma\} = 1 + B_{t_n}.$$

Therefore, by using (7.13), we obtain

$$L := \prod_{n=n_0}^{\infty} L_n \leq e^B < +\infty,$$

which implies that $\{H_k\}$ is well-controlled. □

By employing Theorem 4.1 in conjunction with Lemma 7.3 and our general stability result (Theorem 7.2), we immediately obtain the following stability theorem.

Theorem 7.3 *Let X be a uniformly convex Banach space X with the Opial property. Let $\mathcal{T} \in \mathcal{APNS}(C)$. Consider a generalised Krasnosel'skii-Mann process $gKM(\mathcal{T}, \{c_k\}, \{t_k\})$ such that*

$$\sum_{n=n_0}^{\infty} \sup\{b_{t_n}(z) : z \in \Sigma\} < \infty, \quad (7.14)$$

holds for some $n_0 \in \mathbb{N}$. Assume that the sequence generated by $gKM(\mathcal{T}, \{c_k\}, \{t_k\})$ is an approximate fixed point sequence for every T_s, where $s \in A \subset J$ and A is a generating set for J. Then, $gKM(\mathcal{T}, \{c_k\}, \{t_k\})$ is stable under summable errors.

Similarly, according to Theorem 4.6, we have our next result.

Theorem 7.4 *Let X be a uniformly convex and uniformly smooth Banach space X. Let $\mathcal{T} \in \mathcal{APNS}(C)$. Consider a well-defined generalised Krasnosel'skii-Mann process $gKM_{UB}(\mathcal{T}, \{c_k\}, \{t_k\})$ such that*

$$\sum_{n=n_0}^{\infty} \sup\{b_{t_n}(z) : z \in \Sigma\} < \infty$$

holds for some $n_0 \in \mathbb{N}$. Assume that the sequence generated by $gKM(\mathcal{T}, \{c_k\}, \{t_k\})$ is an approximate fixed point sequence for every T_s. Then, $gKM(\mathcal{T}, \{c_k\}, \{t_k\})$ is stable under summable errors.

Please note that, based on the results discussed above, all specific weak-convergence results from Chap. 4 can be applied similarly, as long as the condition (7.14) is satisfied.

As promised, we need to unpack the assumption $\sum_{n=n_0}^{\infty} \sup\{b_{t_n}(z) : z \in \Sigma\} < \infty$. We will address this through a series of remarks.

Remark 7.3 Firstly, if $\mathcal{T}$ is a nonexpansive semigroup then each $b_{t_n}(x) = 0$ for every $x \in C$ and all n, trivially implying (7.14).

Remark 7.4 The same argument applies to eventually nonexpansive semigroups.

Remark 7.5 When $\mathcal{T}$ is asymptotic nonexpansive semigroup (non-pointwise case) then b_{t_n} is a sequence of numbers (independent of x) which tends to zero as $n \to \infty$ because $a_{t_n} = 1 + b_{t_n} \to 1$. We can modify this process by taking a subsequence $\{t_{n_j}\}$ such that $\sum_{j=1}^{\infty} b_{t_{n_j}} < \infty$.

7.3 Stability of Ishikawa Processes

Remark 7.6 Let us consider a more general case. As we know, $b_{t_n}(x) \to 0$ for every $x \in C$. Assume now that $b_{t_n} \rightrightarrows 0$ on Σ. Then, surely, we can select a subsequence $\{t_{n_j}\}$ such that $\sum_{j=1}^{\infty} \sup\{b_{t_{n_j}}(z) : z \in \Sigma\} < \infty$. This assumption is not unreasonable considering that Σ is actually the set of all results of multiple applications of the generalised Krasnosel'skii-Mann iterative process starting from some of the elements of the sequence $\{x_k\}$, which becomes very close to the sequence of $u_{k+1} = H_k(u_k)$.

7.3 Stability of Ishikawa Processes

In this section, we will illustrate the stability of the generalised Ishikawa iteration processes by applying the general results derived in Sect. 7.1. Please note that the scenario becomes more intricate compared to the Krasnosel'skii-Mann case, necessitating stricter control conditions. This should not come as a surprise, as there is often a typical trade-off between enhanced performance and increased complexity in iterative algorithms.

Lemma 7.4 *Let C be a bounded, closed, and convex subset of a uniformly convex Banach space X, and let $\mathcal{T} \in \mathcal{APNS}(C)$. Consider a well-defined generalised Ishikawa iteration process $gI(\mathcal{T}, \{c_k\}, \{d_k\}, \{t_k\})$. Let for every $k \in \mathbb{N}$, H_k be defined by the formula*

$$H_k(x) = c_k T_{t_k}(d_k T_{t_k}(x) + (1 - d_k)x) + (1 - c_k)x.$$

Denote

$$B_{t_n} = \sup\{b_{t_n}(z) : z \in C\}.$$

Assume that there exists $n_0 \in \mathbb{N}$ such that

$$B := \sum_{n=n_0}^{\infty} B_{t_n} < \infty. \tag{7.15}$$

Then, $\{H_k\}$ is a well-controlled sequence of pointwise Lipschitzian operators.

Proof A straightforward calculation shows that for every $k \in \mathbb{N}$ and all $x, y \in C$,

$$\|H_k(x) - H_k(y)\| \leq A_k(x)\|x - y\| = (1 + D_k(x))\|x - y\|,$$

where

$$D_k(x) = c_k a_{t_k}(d_k T_{t_k}(x) + (1 - d_k)x)(1 + d_k a_{t_k}(x) - d_k) - c_k.$$

Note that $A_k(x) \geq 1$ which follows directly from the fact that $a_{t_k}(x) \geq 1$ for any $x \in C$. This implies that $\{H_k\} \in \mathcal{A}(C)$. By (7.15), there exists a natural number $k_0 \geq n_0$ such that for $k \geq k_0$,

$$\sup\{a_{t_k}(z) : z \in C\} \le 2.$$

Recall that $a_{t_k}(x) = 1 + b_{t_k}(x)$ for every $x \in C$ and that $c_k, d_k \le 1$. Let us fix temporarily $k \ge k_0$, and denote $w = d_k T_{t_k}(x) + (1 - d_k)x$. Calculate

$$\begin{aligned} D_k(x) &\le a_{t_k}(w)(1 + d_k a_{t_k}(x) - d_k) - 1 \\ &= a_{t_k}(w) - 1 + d_k a_{t_k}(w) a_{t_k}(x) - d_k a_{t_k}(w) \\ &= b_{t_k}(w) + d_k a_{t_k}(w)(a_{t_k}(x) - 1) \\ &\le b_{t_k}(w) + 2 b_{t_k}(x). \end{aligned}$$

Therefore, for $k \ge k_0$, we obtain

$$\sup\{D_k(z) : z \in C\} \le 3 \sup\{b_{t_k}(z) : z \in C\} = 3 B_{t_k}.$$

Noting that

$$L_j = \sup\{A_j(x) : x \in C\} \le 1 + \sup\{D_j(x) : x \in C\},$$

we get

$$L = \prod_{j=1}^{\infty} L_j \le e^D < +\infty,$$

where

$$D = \sum_{j=1}^{k_0-1} \sup\{D_k(x) : x \in C\} + 3 \sum_{j=k_0}^{\infty} B_{t_j} < +\infty.$$

We conclude finally that $\{H_k\}$ is a well-controlled sequence of pointwise Lipschitzian operators, as claimed. □

Similarly as in the generalised Krasnosel'skii-Mann process case, we can use Theorems 5.1 and 5.4 (respectively) in conjunction with Lemma 7.4 and our general stability result (Theorem 7.2) to obtain the following stability results for the Ishikawa process.

Theorem 7.5 *Let X be a uniformly convex Banach space X with the Opial property. Let $\mathcal{T} \in \mathcal{APNS}(C)$. Consider a well-defined Ishikawa process $gI(\mathcal{T}, \{c_k\}, \{d_k\}, \{t_k\})$ such that*

$$\sum_{n=n_0}^{\infty} \sup\{b_{t_n}(z) : z \in C\} < \infty \tag{7.16}$$

holds for some $n_0 \in \mathbb{N}$. Assume that the sequence generated by $gI(\mathcal{T}, \{c_k\}, \{d_k\}, \{t_k\})$ is an approximate fixed point sequence for every T_s. Then, $gI(\mathcal{T}, \{c_k\}, \{d_k\}, \{t_k\})$ is stable under summable errors.

Theorem 7.6 *Let X be a uniformly convex and uniformly smooth Banach space X. Consider a well-defined generalised locally uniformly bounded Ishikawa iteration process $gI_{UB}(\mathcal{T}, \{c_k\}, \{d_k\}, \{t_k\})$ such that $d_k \to 0$. Assume that*

$$\sum_{n=n_0}^{\infty} \sup\{b_{t_n}(z) : z \in C\} < \infty,$$

holds for some $n_0 \in \mathbb{N}$. Assume that the sequence generated by this process is an approximate fixed point sequence for every T_s. Then, $gI_{UB}(\mathcal{T}, \{c_k\}, \{d_k\}, \{t_k\})$ is stable under summable errors.

Remark 7.7 Observe that the condition (7.16) is automatically satisfied for nonexpansive and eventually nonexpansive semigroups. Similarly to the Krasnosel'skii-Mann case, if $\mathcal{T}$ is asymptotic nonexpansive semigroup, we can easily adjust the process by selecting a subsequence of $\{t_n\}$ to ensure that (7.16) holds. However, the general case presents more complexity. In the context of the Ishikawa scenario, we must guarantee that $b_{t_n} \rightrightarrows 0$ on the entire set C, rather than just on a smaller subset $\Sigma \subset C$. However, when considering specific examples, it may be sufficient to impose some less stringent assumptions.

7.4 Stability of Implicit Iteration Processes

For simplicity, this section will focus exclusively on the stability of the implicit iterative processes $\{u_k\} = P(C, \mathcal{T}, u_1, \{c_k\}, \{t_k\})$ for asymptotic nonexpansive semigroups. This means that we will always assume that $\mathcal{T} \in \mathcal{ANS}(C)$. To apply the general stability results from Sect. 7.1, we need to represent the implicit iteration processes using asymptotic sequences. We will utilise techniques introduced in the proof of Lemma 6.3, albeit in a different context.

Definition 7.7 The sequence $\{Z_k\} \in \mathcal{A}(C)$, which exists according to Lemma 6.3, will be called an asymptotic nonexpansive sequence associated with the implicit iterative process $\{u_k\} = P(C, \mathcal{T}, u_1, \{c_k\}, \{t_k\})$. Accordingly, $\Gamma(\{Z_k\})$ will be called the iterative process associated with $\{u_k\} = P(C, \mathcal{T}, u_1, \{c_k\}, \{t_k\})$.

Theorem 7.7 *Let X be a uniformly convex Banach space. Additionally, assume that X is either uniformly smooth or it possesses the Opial property. Let C be a bounded, closed and convex subset of X. Let $\mathcal{T}$ be an asymptotic nonexpansive semigroup ($\mathcal{T} \in \mathcal{ANS}(C)$), which is equicontinuous on C. Assume that there exists $\gamma \geq 1$ such that $a_t(x) \leq \gamma$ holds for every $x \in C$ and for all $t \in J = [0, +\infty)$. Let us assume that $\{u_k\} = P(C, \mathcal{T}, u_1, \{c_k\}, \{t_k\})$ is a normalised implicit iteration process. Then, the iteration process $\Gamma(\{Z_k\})$ associated with $\{u_k\} = P(C, \mathcal{T}, u_1, \{c_k\}, \{t_k\})$ is stable under summable errors.*

Proof According to Lemma 6.3, there exists $\{Z_k\} \in \mathcal{A}_c(C)$, which generates the implicit iterative processes $\{u_k\} = P(C, \mathcal{T}, u_1, \{c_k\}, \{t_k\})$ and that the definition of the sequence $\{Z_k\}$ is independent of the selection of the starting point of the process (see Remark 6.2). Recall that $u_{k+1} = Z_k(u_k)$. It can be concluded from Theorem 6.1 regarding the Opial property case and from Theorem 6.2 for the uniformly smooth X that, regardless of the choice of the starting point, the sequence generated by the implicit iterative process converges in the weak topology to a common fixed point of the semigroup $\mathcal{T}$. This means that $\Gamma(\{Z_k\})$ is weak-convergent to $F(\mathcal{T})$ (recall Remark 7.2). In Lemma 6.3, we proved that $\prod_{k=1}^{\infty} A_k < \infty$, which implies that the sequence $\{Z_k\}$ is well-controlled. By employing Theorem 7.2, we conclude that $\Gamma(\{Z_k\})$ is stable under summable errors, as claimed. $\square$

Remark 7.8 By employing Theorem 6.3 alongside [72, Theorem 3.3], one can establish a related result concerning the stability of implicit iteration processes that strongly converge to a common fixed point.

7.5 Notes

Ideas of stability under summable errors for iteration processes can be traced back to the 1967 paper by Ostrowski [88]. Butnariu, Reich, and Zaslavski followed this path, investigating stability of the iterates of one nonexpansive mapping [21, 22], see also works by Reich, Taiwo, and Zaslavski [94, 95, 102, 111], and the literature referred there. Books [109, 110, 112, 113] by Zaslavski provide an excellent exposé of the recent status of the stability theory for iterative processes corresponding to fixed point problems. In particular, [113] discusses cases of summable and non-summable errors stability for the Krasnosel'skii-Mann method.

Most of the material presented in this chapter is loosely based on the recent papers by Kozlowski, [72, 74, 76].

Chapter 8
Semigroups of Monotone Operators

Abstract In this chapter, we explore monotone asymptotic pointwise contractive and nonexpansive semigroups. We demonstrate the existence of common fixed points for these operator semigroups in Banach spaces with a partial order structure. Additionally, we present an algorithm for constructing a common fixed point and analyse the conditions that enable both weak and strong convergence. We also provide illustrative examples and contextualise our findings within the framework of differential equations and dynamical systems theory.

8.1 Introduction

In the last two decades, we have observed the emergence of an interesting research stream focused on fixed point theorems in Banach spaces, or more generally, in metric, hyperbolic, and modular spaces equipped with a partial order. In this context, the Lipschitzian-like assumptions apply only to comparable elements x (in the order sense), unlike in the general theory, where such assumptions are made for all elements. This phenomenon reflects the overarching philosophy that underlies the research presented in this book. We can summarise this philosophy with the assertion that, in many applications, common fixed point results can be established under significantly weaker assumptions than those required by classical fixed point theory, which demands strict nonexpansiveness for all mappings in the semigroup, and across the entire set C.

Since the set of comparable elements may be quite small, proofs of the monotone variants often require the use of innovative techniques, and special attention must be paid to avoid the influence of habit. Generally speaking, all calculations can only be performed on the comparable elements, and to proceed, one must have a foundational point x_0 that is comparable with all elements of the form $T_s(x_0)$. Furthermore, monotone contractions or monotone nonexpansive mappings need not be continuous, whereas the continuity of nonexpansive mappings is typically taken for granted. Moreover, some results may even be counter-intuitive. For example, the uniqueness of a fixed point assumes a new and distinct form. Nevertheless, an interesting and continually evolving body of applications, along with its connections to graph theory, renders this relatively new field both important and promising.

8.2 Preliminaries

In this chapter we assume that the Banach space X is endowed with a partial order "$\preceq$". We say that $x \in X$ and $y \in X$ are comparable if either $x \preceq y$ or $y \preceq x$. Let us recall that an order interval is any of the subsets $[a, \rightarrow) = \{y \in X : a \preceq y\}$ and $(\leftarrow, b] = \{y \in X : y \preceq b\}$. For $a \preceq b$ we define the interval

$$[a, b] = [a, \rightarrow) \cap (\leftarrow, b].$$

We will also assume that the partial order "$\preceq$" and the linear structure of X are linked to each other by

$$a \preceq b, \ c \preceq d \Rightarrow \lambda a + (1 - \lambda)c \preceq \lambda b + (1 - \lambda)d \qquad (8.1)$$

holding for any $\lambda \in [0, 1]$ and all $a, b, c, d \in X$. It immediately follows from (8.1) that

$$a \preceq \lambda a + (1 - \lambda)b \preceq b,$$

whenever $a \preceq b$. This implies that all order intervals are convex. We will also assume that all order intervals are closed.

Remark 8.1 It is well known that intervals within $L^p([0, 1])$ are both convex and closed for any $p > 1$. Recall that, for elements $x, y \in L^p([0, 1])$, the relation $y \preceq x$ is defined by the condition $y(t) \leq x(t)$ almost everywhere on the interval $[0, 1]$. For further details, refer to sources such as [6].

Definition 8.1 We say that $T : C \rightarrow C$ is a monotone (or order-preserving) mapping if the following implication holds,

$$x \preceq y \Rightarrow T(x) \preceq T(y).$$

The common fixed point results for monotone semigroups, along with their proofs, follow similar approaches to our previous non-monotone results from the previous chapters. However, there are important subtleties that require special attention, so we will provide detailed proofs as necessary. One of the major differences is that, to establish our common fixed point existence for the monotone semigroups, we need to establish a connection between the order structure and the semigroup $\mathcal{T}$. This requirement is not surprising; without it, we could encounter a scenario where no element from the set C is comparable to its images under the mappings from

8.2 Preliminaries

$\mathcal{T}$, making it challenging to expect a common fixed point to exist. Therefore, we assume that every mapping T, discussed in this chapter, is monotone and there exists an element $x \in C$ such that $x \preceq T(x)$. Similarly, we will always assume that there is an element $x \in C$ such that $x \preceq T_t(x)$ for all $t \in J$, and we will use this element as the foundation for constructing a common fixed point. Recall that in the previous chapters, we were able to use any $x \in C$ as a starting point.

Remark 8.2 It is worth noting that the assumption $x \preceq T_t(x)$ could be replaced by its reverse, $T_t(x) \preceq x$, and this "reverse theory" would develop in a similar manner. However, for simplicity, we will consistently assume that $x \preceq T_t(x)$ for all $t \in J$.

Let us provide formal definitions of monotone pointwise Lipschitzian mappings and monotone pointwise Lipschitzian semigroups of mappings.

Definition 8.2 Let $T : C \to C$ be a monotone mapping. We say that T is a monotone pointwise Lipschitzian mapping if to every $x \in C$ there exists a number $\alpha(x) \geq 0$ such that such that

$$\|T(x) - T(y)\| \leq \alpha(x)\|x - y\| \tag{8.2}$$

holds for all $y \in C$ such that $y \preceq x$. If $\alpha(x) < 1$ for every $x \in C$ then T will be called a monotone pointwise contraction. Similarly, if $\alpha(x) \leq 1$ for every $x \in C$ then T is called a monotone pointwise nonexpansive mapping.

Note that (8.2) needs to hold only for $y \preceq x$, which implies, among others, that monotone pointwise Lipschitzian mappings do not have to be continuous.

As can be easily inferred, in the "reverse theory", we would need to introduce the "reverse assumption" that (8.2) should hold only for $y \in C$ such that $x \preceq y$. It is important to note that the assumption typically used in the literature is that (8.2) must hold for all comparable x and y. This approach works well for monotone nonexpansive semigroups when $\alpha(x) = 1$ for all $x \in C$. However, in our context, this approach to defining monotone pointwise Lipschitzian mappings is generally too restrictive, as it could unnecessarily complicate or exclude interesting examples; for further details, see Sect. 9.2. Nonetheless, in certain situations, adopting a more restrictive approach is justified—particularly when discussing the convexity of the set of common fixed points. The following definition addresses this scenario.

Definition 8.3 Let $T : C \to C$ be a monotone pointwise Lipschitzian mapping. We say that T has the symmetry property if the condition (8.2) holds for all $y \in C$ that are comparable with x.

By replacing the phrase "pointwise Lipschitzian mapping" with "monotone pointwise Lipschitzian mapping" in the semigroup Definitions 1.5 through 1.10, we derive the monotone variants of all the semigroups of interest. In particular, the class of all monotone pointwise Lipschitzian semigroups replaces the class of all pointwise Lipschitzian semigroups. We will add the prefix "m" to the class definitions of these semigroups. For instance, a semigroup $\mathcal{T}$ will be termed monotone asymptotic pointwise contractive if each $T \in \mathcal{T}$ is a monotone pointwise Lipschitzian mapping and

$\limsup_{t\to\infty} \alpha_t(x) < 1$ for every $x \in C$. The class of all monotone asymptotic pointwise contractive semigroups on C is denoted by $m\mathcal{APCS}(C)$. Similarly, $\mathcal{T}$ is called monotone asymptotic pointwise nonexpansive if each $T \in \mathcal{T}$ is a monotone pointwise Lipschitzian mapping and $\limsup_{t\to\infty} \alpha_t(x) \leq 1$ for every $x \in C$. The class of all monotone asymptotic pointwise nonexpansive semigroups on C is denoted by $m\mathcal{APNS}(C)$.

As mentioned previously, many classical spaces, including Hilbert space and Banach spaces l^p with $1 < p < \infty$, possess the Opial property. However, important uniformly convex spaces like $L^p[0, 1]$ (for $1 < p < \infty$) do not satisfy this condition when $p \neq 2$. In the context of monotone mappings, the following concept can provide a remedy.

Definition 8.4 ([3]) We say that $(X, \|\cdot\|, \preceq)$ have the monotone Opial property if for each nondecreasing (resp. nonincreasing) sequence $\{x_n\}$ of elements of X such that $x_n \rightharpoonup x$, and for any $y \in X$ such that $y \neq x$,

$$\liminf_{n\to\infty} \|x_n - x\| < \liminf_{n\to\infty} \|x_n - y\|,$$

provided that $x_n \preceq y$ (resp. $y \preceq x_n$) for every $n \in \mathbb{N}$.

Clearly, the Opial property implies the monotone Opial property; however, it is important to note that these two properties are not equivalent.

Theorem 8.1 ([3, Theorem 3.9]) *Let $(X, \|\cdot\|, \preceq)$ be a uniformly convex, partially ordered Banach space for which order intervals are convex and closed. Assume that the norm $\|\cdot\|$ is monotone. Then X has the monotone Opial property.*

As an important example, we deduce from Theorem 8.1, that all spaces $L^p[0, 1]$ possess the monotone Opial property, where $p > 1$.

Let us recall that $\|\cdot\|$ is said to be monotone if $u \preceq v \preceq w$ implies that

$$\max\{\|v - u\|, \|w - v\|\} \leq \|w - u\|$$

for any $u, v, w \in X$. Note that by taking $u = 0$, we deduce that $\|v\| \leq \|w\|$ whenever $0 \preceq v \preceq w$. It is also worthwhile noticing that if the norm is monotone and $\{x_n\}$ is nondecreasing (resp. nonincreasing) then the sequence $\{\|x_n - y\|\}$ is nonincreasing for any y such that $x_n \preceq y$ (resp. $y \preceq x_n$) for $n \in \mathbb{N}$.

8.3 Existence of Common Fixed Points

Let us start with the following technical results.

Lemma 8.1 *Let $C \subset X$ be a convex and weakly compact set, and let $\mathcal{T}$ be a monotone pointwise Lipschitzian semigroup on C. Additionally, assume that there exists $x \in C$ such that $x \preceq T_t(x)$ for all $t \in J$. Define*

8.3 Existence of Common Fixed Points

$$C_x = C \cap \left(\bigcap_{s \in J} [T_s(x), \rightarrow) \right). \tag{8.3}$$

Then, C_x is nonempty, closed, convex, and weakly compact. Furthermore, it follows that $T_s(C_x) \subset C_x$ for every $s \in J$.

Proof Observe that for any finite sequence $\{t_1, \ldots, t_n\}$ of elements from J,

$$T_{t_i}(x) \preceq T_{t_1 + \cdots + t_n}(x),$$

because, for any $s \in J$, $x \preceq T_s(x)$ and T_s is monotone. Therefore,

$$T_{t_1 + \cdots + t_n}(x) \in C \cap \left(\bigcap_{s \in \{t_1, \ldots, t_n\}} [T_s(x), \rightarrow) \right) \subset C \cap \left(\bigcap_{s \in J} [T_s(x), \rightarrow) \right) = C_x,$$

which proves that C_x is not empty. Since C is closed and convex, and each interval of the form $[T_s(x), \rightarrow)$ shares these properties, it is evident that C_x is also closed and convex. Consequently, as C is weakly compact, we can conclude that C_x is weakly compact as well. It remains to prove that $T_s(C_x) \subset C_x$ for every $s \in J$. Fix $y \in C_x$ and $s \in J$. By the definition of C_x, we have $T_t(x) \preceq y$ for all $t \in J$. This implies that, for a given $t \in J$, we have $T_{t+s}(x) \preceq T_s(y)$, since T_s is monotone. On the other hand, if follows from $x \preceq T_s(x)$ that $T_t(x) \preceq T_{t+s}(x)$. Finally, we conclude from the above that $T_t(x) \preceq T_s(y)$ for every $t \in J$. This implies that $T_s(y) \in C_x$, as claimed. □

The following result is a monotone variant of the existence theorem for asymptotic pointwise contractive semigroups, thus linking it to Theorem 2.1.

Theorem 8.2 *Assume that X is a uniformly convex Banach space. Let $C \subset X$ be a nonempty, bounded, closed and convex set, and let $\mathcal{T}$ be a monotone asymptotic pointwise contractive semigroup on C. Assume in addition that there exists $x \in C$ such that $x \preceq T_t(x)$ for all $t \in J$. Then, there exists a common fixed point $z \in Fix(\mathcal{T})$ such that $x \preceq z$ and $\|T_t(x) - z\| \to 0$ as $t \to \infty$. Moreover, if w is another common fixed point, which is comparable with z, then $w = z$.*

Proof Let $C_x \subset C$ be defined as in (8.3) of Lemma 8.1. From this lemma it follows that C_x is nonempty, closed, convex, and weakly compact, as well as that $T_s(C_x) \subset C_x$ for every $s \in J$. Define the type function $\varphi : C_x \to [0, +\infty)$ by

$$\varphi(y) = \limsup_{t \to \infty} \|T_t(x) - y\|,$$

for any $y \in C_x$. As clarified in Remark 2.2, since φ is weakly lower semicontinuous on a weakly compact set C_x then it attains its minimum at a point $z \in C_x$. Since $z \in C_x$ then $T_t(x) \preceq z$ for any $t \in J$. Observe that,

$$\varphi(T_s(z)) = \limsup_{t\to\infty} \|T_t(x) - T_s(z)\| = \limsup_{t\to\infty} \|T_{s+t}(x) - T_s(z)\| \quad (8.4)$$
$$\leq \limsup_{t\to\infty} \alpha_s(z)\|T_t(x) - z\| = \alpha_s(z)\varphi(z),$$

holds for all $s \in J$. Note that in (8.4), we used the inequality $T_{s+t}(x) \preceq T_s(z)$, which arises from $T_t(x) \preceq z$ due to the monotonicity of T_s. By leveraging the minimality of φ at z, we infer from (8.4) that

$$\varphi(z) \leq \varphi(T_s(z)) \leq \alpha_s(z)\varphi(z). \quad (8.5)$$

By passing with s to infinity in (8.5) we obtain

$$\varphi(z) \leq \alpha(z)\varphi(z),$$

where $\alpha(z) = \limsup_{t\to\infty} \alpha_t(z) < 1$, which implies that $\varphi(z) = 0$. Then, by (8.4) again, for every $s \in J$,

$$\lim_{t\to\infty} \|T_t(x) - T_s(z)\| = 0. \quad (8.6)$$

By applying (8.6) with $s = 0$, we conclude that

$$\lim_{t\to\infty} \|T_t(x) - z\| = 0. \quad (8.7)$$

From (8.6) and (8.7) it follows immediately that $T_s(z) = z$. Since $s \in J$ was chosen arbitrarily, it follows that $z \in F(\mathcal{T})$, as claimed.

To prove the uniqueness among comparable to z common fixed points, let us take another $w \in F(\mathcal{T})$ which is comparable with z. Assume that $w \preceq z$ (the proof for the reverse case is analogous). Therefore, for each $t \in J$,

$$\|z - w\| = \|T_t(z) - T_t(w)\| \leq \alpha_t(z)\|z - w\|. \quad (8.8)$$

By passing with t to infinity in (8.8), we obtain

$$\|z - w\| \leq \alpha(z)\|z - w\|.$$

Since $\alpha(z) < 1$, this inequality implies that $w = z$. The proof of the theorem is complete. □

We will need the following monotone variant of the definition of a closed set.

Definition 8.5 We say that the set $D \subset X$ is monotonically closed if for any sequence $\{x_n\}$ of elements from D and an element $x \in X$ such that $x_n \preceq x$ for every $n \in \mathbb{N}$, the convergence $x_n \to x$ implies that $x \in D$.

In our next result, we will prove the existence of a common fixed point for monotone asymptotic pointwise nonexpansive semigroups. The existence of this fixed

8.3 Existence of Common Fixed Points

point can be directly derived from the Knaster-Tarski Theorem; for more details, see Sect. 8.6 "Notes". However, the proof presented below follows the general techniques used in the previous chapters and does not rely on external results. Therefore, for the sake of completeness, it is presented here in its entirety.

Theorem 8.3 *Assume that X is a uniformly convex space and let $\mathcal{T}$ be a monotone asymptotic pointwise nonexpansive semigroup defined on a nonempty, bounded, closed, and convex subset C of X. Additionally, suppose there exists an element $x \in C$ such that $x \preceq T_t(x)$ for all $t \in J$, where each T_t is continuous on the subset C_x. Under these conditions, there exists a common fixed point $z \in F(\mathcal{T})$ such that $x \preceq z$. Furthermore, $F(\mathcal{T})$ is monotonically closed. Notably, if $\mathcal{T}$ is a monotone nonexpansive semigroup, the requirement for each T_t to be continuous can be omitted.*

Proof Without any loss of generality, we can assume that each $a_t(z) \geq 1$, where $\|T(z) - T(y)\| \leq a_t(z)\|z - y\|$ for any $y \preceq z$. Recall that

$$C_x = C \cap \left(\bigcap_{s \in J} [T_s(x), \rightarrow) \right).$$

In the proof of Theorem 8.2, we demonstrated that C_x is a nonempty, convex, and weakly compact subset of C. Define the type function $\tau : C_x \rightarrow [0, +\infty)$ by

$$\tau(y) = \limsup_{t \to \infty} \|T_t(x) - y\|^2.$$

Similarly to the proof of Theorem 8.2, we can deduce that τ attains its minimum at an element $z \in C_x$. Let us fix arbitrarily $t, s, u \in J$. Because X is uniformly convex it follows from Proposition 2.1 that for each $d > 0$ there exists a continuous, strictly increasing and convex function $\lambda : [0, \infty) \rightarrow [0, \infty)$ such that $\lambda(t) = 0$ if and only if $t = 0$, and

$$\|cw + (1-c)v\|^2 \leq c\|w\|^2 + (1-c)\|v\|^2 - c(1-c)\lambda(\|w - v\|), \quad (8.9)$$

holds for any $c \in [0, 1]$ and all $w, v \in X$ such that $\|w\| \leq d$ and $\|v\| \leq d$. Applying (8.9) to $w = T_{s+u+t}(x) - T_s(z)$, $v = T_{s+u+t}(x) - T_u(z)$, $d = \text{diam}(C)$ and $c = \frac{1}{2}$ we obtain the inequality

$$\left\| T_{s+u+t}(x) - \frac{1}{2}\left(T_s(z) + T_u(z)\right) \right\|^2 \quad (8.10)$$

$$\leq \frac{1}{2}\left\| T_{s+u+t}(x) - T_s(z) \right\|^2 + \frac{1}{2}\left\| T_{s+u+t}(x) - T_u(z) \right\|^2 - \frac{1}{4}\lambda(\|T_s(z) - T_u(z)\|).$$

Since $z \in C_x$, $T_p(x) \preceq z$ for any $p \in J$, and hence, it follows from (8.10) that

$$\left\| T_{s+u+t}(x) - \frac{1}{2}\big(T_s(z) + T_u(z)\big) \right\|^2 \tag{8.11}$$

$$\leq \frac{1}{2} a_s(z)^2 \|T_{u+t}(x) - z\|^2 + \frac{1}{2} a_u(z)^2 \|T_{s+t}(x) - z\|^2 - \frac{1}{4}\lambda(\|T_s(z) - T_u(z)\|).$$

By letting $t \to \infty$ on both sides of (8.11) and utilising the minimality of τ at z, we obtain

$$\tau(z) \leq \tau\left(\frac{T_s(z) + T_u(z)}{2}\right) \leq \left(\frac{1}{2} a_s(z)^2 + \frac{1}{2} a_u(z)^2\right)\tau(z) - \frac{1}{4}\lambda(\|T_s(z) - T_u(z)\|),$$

from which, by setting $u = 0$, we infer that

$$0 \leq \frac{1}{2}\lambda(\|T_s(z) - z\|) \leq (a_s(z)^2 - 1)\tau(z). \tag{8.12}$$

If $\mathcal{T}$ is monotone nonexpansive then each $a_s(z) = 1$. This implies that for every $s \in J$, $\lambda(\|T_s(z) - z\|) = 0$ and, consequently, that $T_s(z) = z$. If $\mathcal{T}$ is not monotone nonexpansive, we will have to use the continuity argument. Since $a_s(z) \to 1$ as $s \to \infty$, it follows from (8.12) that $\lambda(\|T_s(z) - z\|) \to 0$. This implies that

$$\lim_{s \to \infty} \|T_s(z) - z\| = 0, \tag{8.13}$$

since the continuous function λ is strictly increasing and takes value equal to zero only at zero. Clearly, this implies that

$$\lim_{s \to \infty} \|T_{t+s}(z) - z\| = 0, \tag{8.14}$$

for every $t \in J$. On the other hand, by the continuity of T_t it follows from (8.13) that

$$\lim_{s \to \infty} \|T_{t+s}(z) - T_t(z)\| = \lim_{s \to \infty} \|T_t(T_s(z)) - T_t(z)\| = 0. \tag{8.15}$$

By the uniqueness of the limit, it follows from (8.14) and (8.15) that $T_t(z) = z$. Given that $s \in J$ was chosen arbitrarily, this implies $z \in F(\mathcal{T})$. It is evident that $x \preceq z$ because $z \in C_x$.

To prove that $F(\mathcal{T})$ is monotonically closed, it suffices to demonstrate that every $F(T_t)$ is monotonically closed. To this end, fix $t \in J$, let $v_n \in F(T_t)$, $v_n \preceq v$ for every $n \in \mathbb{N}$, and let $v_n \to v$. Note that

$$\|T_t(v) - v\| \leq \|T_t(v) - v_n\| + \|v_n - v\| = \|T_t(v) - T_t(v_n)\| + \|v_n - v\|$$
$$\leq \alpha_t(v)\|v - v_n\| + \|v_n - v\| \to 0,$$

as $n \to \infty$. This completes the proof. $\qquad\square$

It is important for the reader to recall that, in the non-monotone case, Theorem 2.3 also states that $F(\mathcal{T})$ is convex—a property absent from Theorem 8.3. This situation is not surprising, given the lack of relationship between elements that are not comparable. However, a refined form of convexity can be achieved under additional assumptions.

Theorem 8.4 *Suppose all assumptions of Theorem 8.3 are satisfied. Additionally, assume each $T_t \in \mathcal{T}$ possesses the symmetry property. Let $f_1, f_2 \in F(\mathcal{T})$ be comparable and let $c \in [0, 1]$. Then $f := cf_1 + (1 - c)f_2 \in F(\mathcal{T})$.*

Proof The proof requires a monotone version of Lemma 2.1. By inspecting the proof of this lemma, it is easy to see that, thanks to the assumed symmetry property, $\|u\|, \|v\| \leq 1$, where u, v are defined as in (2.15). The rest of the proof of the lemma remains unchanged. Then, we can follow the proof of convexity of $F(\mathcal{T})$ as presented in Theorem 2.3 remembering to apply this variant of the lemma as well as the monotone pointwise Lipschitzian property of each T_t to conclude that

$$\|T_t(f) - f\| \to 0.$$

By using the same continuity argument as in the proof of Theorem 8.3, we conclude that $T_t(f) = f$ for every $t \in J$. □

8.4 Weak-Convergence

In this section, we always assume that C is a nonempty, bounded, closed, and convex subset of a uniformly convex Banach space X. We will also assume that J is an additive semigroup of nonnegative integers denoted by $\mathbb{N}_0$. Based on our discussion in Remark 3.2, we note that the semigroup J is finitely generated. This means that every number in J can be expressed as the sum of nonnegative multiples of elements belonging to a finite set $G = \{\alpha_1, ..., \alpha_M\}$, where each $\alpha_i \in \mathbb{N}$. Furthermore, in this case, to demonstrate that $w \in \mathcal{T}$, it suffices to show that $T_s(w) = w$ for each $s \in G$. We can summarise this symbolically by

$$\bigcap_{s \in G} F(T_s) = F(\mathcal{T}). \tag{8.16}$$

In our main result of this section, Theorem 8.6, we assume that $J \subset \mathbb{N}_0$ and thus it is finitely generated. While this is undoubtedly a significant limitation, it simplifies our proceedings and enables us to eliminate several artificial assumptions. It is worth mentioning that, due to the nature of monotone nonexpansive settings, we typically deal with very small subsets of X, which are connected in a chain with the starting point x_0. From this perspective, and considering typical applications discussed in Sects. 8.6 and 9.2, these limiting assumptions are not unusual.

To derive our weak convergence results, we will adapt to our setting the following iteration process, commonly referred to in the literature as the Krasnoselskii-Ishikawa iteration process, [45, 78]. We will observe that this process is well-suited to the demands of monotone pointwise asymptotic nonexpansive semigroups.

Definition 8.6 Let $\mathcal{T} \in m\mathcal{APNS}(C)$. Let us fix $\lambda \in (0,1)$. Let $\{t_n\}$ be a sequence of elements in J. The monotone Krasnoselskii-Ishikawa iteration process $KIS(\mathcal{T}, \lambda, \{t_n\}, x_0)$ is defined by the formula

$$x_{n+1} = (1-\lambda)x_n + \lambda T_{t_n}(x_n), \tag{8.17}$$

where $x_0 \in C$ is the starting element chosen so that $x_0 \preceq T_t(x_0)$ for all $t \in J$. Assume that each T_t is continuous on C_{x_0}. For technical reasons we will assume that the sequence has been chosen in so that

$$t_n \geq \sum_{i=0}^{n-1} t_i, \tag{8.18}$$

and

$$\sum_{n=1}^{\infty} b_{t_n}(x) < \infty \tag{8.19}$$

holds for every $x \in C$.

Proposition 8.1 *The sequence $\{x_n\}$ defined in (8.17) is nondecreasing. Moreover, for every $n \in \mathbb{N}$,*

$$x_n \preceq T_s(x_n)$$

hold for $s \geq \sum_{i=0}^{n-1} t_i$. In particular, in view of (8.18),

$$x_n \preceq T_{t_n}(x_n). \tag{8.20}$$

Proof Recall first that $x_0 \preceq T_{t_0}(x_0)$ which, by (8.17) and the properties of $\preceq$, implies that $x_0 \preceq x_1$. Define

$$D_0 = C \cap \left(\bigcap_{s \geq t_0} (\leftarrow, T_s(x_0)] \right)$$

and

$$D_1 = C \cap \left(\bigcap_{s \geq t_0 + t_1} (\leftarrow, T_s(x_0)] \right)$$

and observe that each of these sets in nonempty (since each one contains x_0), convex and closed. Clearly $D_0 \subset D_1$. Let $s \geq t_0$. Then $T_{t_0}(x_0) \preceq T_s(x_0)$ since $x_0 \preceq T_{s-t_0}(x_0)$ and T_{t_0} is monotone. This implies that $T_{t_0}(x_0) \in D_0$ and, consequently, by the convexity of D_0, that $x_1 \in D_0$. This, together with the already established $x_0 \preceq x_1$, implies that for every $s \geq t_0$ we have

8.4 Weak-Convergence

$$x_1 \preceq T_s(x_0) \preceq T_s(x_1),$$

and in particular that $x_1 \preceq T_{t_1}(x_1)$. Therefore,

$$x_1 = (1 - \lambda)x_1 + \lambda x_1 \preceq (1 - \lambda)x_1 + \lambda T_{t_1}(x_1) = x_2.$$

Using the same argument as before but with D_1 instead of D_0 we conclude that $x_2 \in D_1$. Consequently, $x_2 \preceq T_s(x_2)$ for $s \geq t_0 + t_1$. Continuing this procedure by the mathematical induction we conclude that $\{x_n\}$ is nondecreasing and that $x_n \preceq T_s(x_n)$ for $s \geq \sum_{i=0}^{n-1} t_i$, as claimed. □

Corollary 8.1 *According to Proposition 8.1, the sequence $\{x_n\}$ generated by the monotone Krasnoselskii-Ishikawa iteration process $KIS(\mathcal{T}, \lambda, \{t_n\}, x_0)$, is nondecreasing. Coupled with the monotonicity of each $T_p \in \mathcal{T}$, this implies that elements $T_p(x_k)$ and $T_p(x_m)$ are comparable for any $p \in J$ and any $k, m \in \{0, 1, \ldots\}$.*

Given $KIS(\mathcal{T}, \lambda, \{t_n\}, x_0)$, we will frequently utilise the set $K \subset C$ defined by

$$K = \{y \in C : x_n \preceq y, \; n = 0, 1, 2 \ldots\} = \bigcap_{n=0}^{\infty} K_n,$$

where $K_n = C \cap [x_n, \rightarrow)$. Note that $\{K_n\}$ is a nonincreasing (in the inclusion sense) sequence of nonempty, bounded, closed, and convex subsets of the uniformly convex Banach space X. This guarantees that $K = \bigcap_{n=0}^{\infty} K_n \neq \emptyset$.

Remark 8.3 Observe that if $w_n \in K_n$ for every n and $w_n \rightharpoonup w$, then $w \in K$. Indeed, by Mazur's Theorem, each K_n is not only norm closed but also weak closed. Therefore, it follows immediately that $w \in K_n$ for every n, since $\{w_k\}_{k \geq n} \subset K_n$.

Lemma 8.2 *Let C be a nonempty, bounded, closed, and convex subset of a uniformly convex Banach space X. Assume that $KIS(\mathcal{T}, \lambda, \{t_n\}, x_0)$ is the monotone Krasnoselskii-Ishikawa iteration process. Let $w \in C$ be a common fixed point for a monotone nonexpansive semigroup $\mathcal{T}$ such that $x_0 \preceq w$. Then, $w \in K$ and there exists $r \in \mathbb{R}$ such that*

$$\lim_{k \to \infty} \|x_k - w\| = r. \tag{8.21}$$

Proof First, let us prove by induction that $w \in K$. Since $x_0 \preceq w$, it follows that $w \in K_0$. It remains to prove that if $w \in K_n$ then $w \in K_{n+1}$. Since, by (8.20), $x_n \preceq T_{t_n}(x_n)$ then $x_{n+1} \in [x_n, T_{t_n}(x_n)]$ as intervals are convex. This implies that

$$x_{n+1} \preceq T_{t_n}(x_n) \preceq T_{t_n}(w) = w,$$

where we applied the inductive assumption $x_n \preceq w$ along with the monotonicity of the mapping T_{t_n}. Consequently, we conclude that $w \in K_{n+1}$, thereby completing the inductive step and demonstrating that, in fact, $w \in K$. Define the sequence of mappings $H_n : C \to C$ defined by $H_n(u) = (1 - \lambda)u + \lambda T_{t_n}(u)$. It is easy to verify that the $\{H_n\}$ is an asymptotic pointwise sequence with $A_n(x) = \max\{1, \lambda a_{t_n}(x)\}$, and that $x_{n+1} = H_n(x_n)$. Moreover, it follows from the assumption (8.19) that $\sum_{n=1}^{\infty} B_n(w) < \infty$, where $B_n(x) = A_n(x) - 1$. By repeating the calculation from Lemma 3.5, while remembering that every $x_n \preceq w$, we can affirm that the limit (8.21) exists. $\square$

The next result is a monotone Krasnosel'skii-Ishikawa analogue of Lemma 4.2 related to the Krasnosel'skii-Mann process.

Lemma 8.3 *Let C be a nonempty, bounded, closed, and convex subset of a uniformly convex Banach space X. Assume that $KIS(\mathcal{T}, \lambda, \{t_n\}, x_0)$ is the monotone Krasnoselskii-Ishikawa iteration process. Let $\{x_k\}$ be a sequence generated by (8.17). Then*

$$\lim_{k \to \infty} \|T_{t_k}(x_k) - x_k\| = 0 \qquad (8.22)$$

and

$$\lim_{k \to \infty} \|x_{k+1} - x_k\| = 0.$$

Proof It follows from Theorem 8.3, that there exists a common fixed point w such that $x_0 \preceq w$. From Lemma 8.2 we know that $w \in K$ and there exists an $r \in \mathbb{R}$ such that $\lim_{k \to \infty} \|x_k - w\| = r$. Since each $x_k \preceq w$, it follows that

$$\limsup_{k \to \infty} \|T_{t_k}(x_k) - w\| = \limsup_{k \to \infty} \|T_{t_k}(x_k) - T_{t_k}(w)\| \leq \limsup_{k \to \infty} a_{t_k}(w)\|x_k - w\| = r.$$

Observe that

$$\lim_{k \to \infty} \|\lambda(T_{t_k}(x_k) - w) + (1 - \lambda)(x_k - w)\| = \lim_{k \to \infty} \|x_{k+1} - w\| = r.$$

By Lemma 3.3 applied to $u_k = x_k - w$, $v_k = T_{t_k}(x_k) - w$, we obtain

$$\lim_{k \to \infty} \|T_{t_k}(x_k) - x_k\| = 0,$$

which by the construction of the sequence $\{x_k\}$ is equivalent to

$$\lim_{k \to \infty} \|x_{k+1} - x_k\| = 0.$$

The proof is complete. $\square$

Just as we did for the non-monotone generalised Kransosel'skii-Manna and Ishikawa processes, we will now introduce the concept of being well-defined for the monotone Krasnoselskii-Ishikawa iteration processes.

8.4 Weak-Convergence

Definition 8.7 We say that a monotone Krasnosel'skii-Ishikawa iteration process $KIS(\mathcal{T}, \lambda, \{t_n\}, x_0)$ is well-defined if the sequence $\{x_k\}$, generated by it, is regular with respect to $\mathcal{T}$, and for any $s \in J$,

$$\lim_{k \to \infty} a_{s+t_k}(x_k) = 1.$$

By making additional assumptions about the norm in X, we can now establish a result that serves a similar purpose to the Demiclosedness Principle within our monotone theory. We can readily identify several arguments that are analogous to those in our previously established variants of the Demiclosedness Principle. However, the necessity of demonstrating required comparability between elements before employing any non-expansiveness type arguments, general monotonically related arguments, and the specifics of the monotone Krasnoselskii-Ishikawa iteration process require careful attention.

Theorem 8.5 *Let C be a nonempty, bounded, closed, and convex subset of a uniformly convex Banach space X. In addition, let us assume that the norm in X is monotone and has the monotone Opial property. Assume that $KIS(\mathcal{T}, \lambda, \{t_n\}, x_0)$ is the monotone Krasnoselskii-Ishikawa iteration process, where $\mathcal{T} \in m\mathcal{APNS}(C)$ and $J \subset \mathbb{N}_0$ is finitely generated by $G = \{\alpha_1, \ldots, \alpha_M\}$. Assume that T_s is continuous for each $s \in G$. Let $\{x_k\}$ be a sequence generated by $KIS(\mathcal{T}, \lambda, \{t_n\}, x_0)$. Assume there exists $w \in C$ and a subsequence $\{y_k\} = \{x_{n_k}\}$ such that $y_k \rightharpoonup w$. If for every $s \in G$,*

$$\|T_s(y_k) - y_k\| \to 0 \qquad (8.23)$$

then $w \in F(\mathcal{T})$.

Proof It follows from Remark 8.3 that $w \in K$. Let us define $\varphi : K \to \mathbb{R}^+$ by

$$\varphi(x) = \limsup_{k \to \infty} \|y_k - x\|.$$

Since $w \in K$, it follows that $y_k \preceq w$ for every $k = 0, 1, \ldots$. Using this comparability along with the formula (8.23), we conclude that

$$\varphi(T_{ms}(w)) \leq \limsup_{k \to \infty} \|T_{ms}(y_k) - T_{ms}(w)\| + \limsup_{k \to \infty} \|T_{ms}(y_k) - y_k\| \quad (8.24)$$
$$\leq \limsup_{k \to \infty} a_{ms}(w)\|y_k - w\| = a_{ms}(w)\varphi(w)$$

holds for any $s \in J$ and any $m \in \mathbb{N}$. Keeping in mind that $\lim_{m \to \infty} a_{ms}(w) = 1$, take the limits of both sides as $m \to \infty$ to obtain

$$\lim_{m \to \infty} \varphi(T_{ms}(w)) \leq \varphi(w).$$

On the other hand, by the monotone weak-Opial property for any $x \in K$ different from w we have

$$\varphi(w) = \limsup_{k\to\infty} \|y_k - w\| < \limsup_{k\to\infty} \|y_k - x\| = \varphi(x),$$

which implies that

$$\varphi(w) = \inf\{\varphi(x) : x \in K\}.$$

This, together with (8.24), gives us

$$\lim_{m\to\infty} \varphi(T_{ms}(w)) = \varphi(w).$$

Proceeding exactly like in Theorem 3.3 we can prove that

$$\lim_{m\to\infty} \lambda(\|T_{ms}(w) - w\|) = 0,$$

where the function λ is defined in Proposition 2.1. Using the properties of λ, we infer that $T_{ms}(w) \to w$ as $m \to \infty$. By the assumed continuity of T_s, we infer from this that

$$T_s(T_{ms}(w)) \to T_s(w), \qquad (8.25)$$

when $m \to \infty$. Using the same argument as above, we conclude that $T_{(1+m)s}(w) \to w$ as $m \to \infty$. Thus,

$$T_s(T_{ms}(w)) = T_{(1+m)s}(w) \to w. \qquad (8.26)$$

Given the uniqueness of the limit, the Eqs. (8.25) and (8.26) establish that $T_s(w) = w$ for every $s \in G$. Use the equality (8.16) to conclude that $w \in F(\mathcal{T})$. The proof is complete. $\square$

Lemma 8.4 *Let C be a nonempty, bounded, closed, and convex subset of a uniformly convex Banach space X. Let $KIS(\mathcal{T}, \lambda, \{t_n\}, x_0)$ be the monotone Krasnoselskii-Ishikawa iteration process finitely generated by $G = \{\alpha_1, \ldots, \alpha_M\}$. Denote by $\{x_k\}$ the sequence generated by this process. Assume that for every $s \in G$ there exists a strictly increasing, quasi-periodic sequence of natural numbers $\{j_k : k = 1, 2, \ldots\}$ with a quasi-period p_s such that $t_{j_{k+1}} = s + t_{j_k}$. Then*

$$\lim_{k\to\infty} \|x_k - x_{j_k}\| = 0, \qquad (8.27)$$

and for every $s \in G$

$$\lim_{k\to\infty} \|T_s(x_k) - x_k\| = 0. \qquad (8.28)$$

Proof Let us fix $s \in G$. Observe that by the quasi-periodicity of $\{j_k\}$, for every positive integer k we have $|k - j_k| \le p_s$. Assume that $k - s \le j_k \le k$ (the proof for the other case is identical). Fix $\varepsilon > 0$. Note that by Lemma 8.3, $\|x_{k+1} - x_k\| < \dfrac{\varepsilon}{p_s}$ for k sufficiently large. Therefore, for k sufficiently large,

8.4 Weak-Convergence

$$\|x_k - x_{j_k}\| \le \|x_k - x_{k-1}\| + \cdots + \|x_{j_k+1} - x_{j_k}\| \le p_s \frac{\varepsilon}{p_s} = \varepsilon,$$

proving (8.27). Note that

$$\|x_{j_k} - x_{j_{k+1}}\| \to 0 \text{ as } k \to \infty.$$

Indeed, it follows from (8.27) and from Lemma 8.3 that

$$\|x_{j_k} - x_{j_{k+1}}\| \le \|x_{j_k} - x_k\| + \|x_k - x_{k+1}\| + \|x_{k+1} - x_{j_{k+1}}\| \to 0. \qquad (8.29)$$

Note that

$$\|x_{j_k} - T_s(x_{j_k})\| \to 0 \text{ as } k \to \infty. \qquad (8.30)$$

To prove (8.30), observe that

$$\begin{aligned}
\|x_{j_k} - T_s(x_{j_k})\| &\le \|x_{j_k} - x_{j_{k+1}}\| + \|x_{j_{k+1}} - T_{t_{j_{k+1}}}(x_{j_{k+1}})\| \\
&\quad + \|T_{t_{j_{k+1}}}(x_{j_{k+1}}) - T_{t_{j_{k+1}}}(x_{j_k})\| \\
&\quad + \|T_{t_{j_{k+1}}}(x_{j_k}) - T_{s+t_{j_k}}(x_{j_k})\| + \|T_{s+t_{j_k}}(x_{j_k}) - T_s(x_{j_k})\| \\
&\le \|x_{j_k} - x_{j_{k+1}}\| + \|x_{j_{k+1}} - T_{t_{j_{k+1}}}(x_{j_{k+1}})\| + a_{t_{j_{k+1}}}(x_{j_{k+1}})\|x_{j_{k+1}} \\
&\quad - x_{j_k}\| + a_{s+t_{j_k}}(x_{j_k}))\|T_{t_{j_{k+1}} - t_{j_k} - s}(x_{j_k}) \\
&\quad - x_{j_k}\| + a_s(x_{j_k})\|T_{t_{j_k}}(x_{j_k}) - x_{j_k}\|,
\end{aligned}$$

which tends to zero by (8.29), because of (8.22) in Lemma 8.3, by the assumption that $t_{j_{k+1}} - t_{j_k} - s = 0$ for $s \in G$, and by the assumption that the process is well-defined, therefore, that the sequence $\{x_k\}$ is regular. Recall that the latter assertions imply that $a_{t_{j_{k+1}}}(x_{j_{k+1}}) \to 1$, $a_{s+t_{j_k}}(x_{j_k})) \to 1$, and $\sup\{a_s(x_{j_k}) : k \in \mathbb{N}\} < +\infty$. It is important to emphasise that we utilised the inequalities $T_p(x_{j_k}) \preceq T_p(x_{j_{k+1}})$ and $T_p(x_{j_k}) \preceq T_p(x_k)$, which hold for any $p \in J$, as a consequence of the conditions $x_{j_k} \preceq x_{j_{k+1}}$ and $x_{j_k} \preceq x_k$.

Similarly,

$$\begin{aligned}
\|x_k - T_s(x_k)\| &\le \|x_k - x_{j_k}\| + \|x_{j_k} - T_{t_{j_k}}(x_{j_k})\| + \|T_{t_{j_k}}(x_{j_k}) - T_{s+t_{j_k}}(x_{j_k})\| \\
&\quad + \|T_{s+t_{j_k}}(x_{j_k}) - T_s(x_{j_k})\| + \|T_s(x_{j_k}) - T_s(x_k)\| \\
&\le \|x_k - x_{j_k}\| + \|x_{j_k} - T_{t_{j_k}}(x_{j_k})\| + a_{t_{j_k}}(x_{j_k})\|x_{j_k} - T_s(x_{j_k})\| \\
&\quad + a_s(x_{j_k}))\|T_{t_{j_k}}(x_{j_k}) - x_{j_k}\| + a_s(x_k)\|x_{j_k} - x_k\|
\end{aligned}$$

tends to the zero as $k \to \infty$, by (8.29), (8.22), (8.30), and by the assumption that the process is well-defined. Therefore, the proof of (8.28) and of the lemma is complete. □

We are now ready to present the main result of this section, the weak convergence theorem for the monotone Krasnoselskii-Ishikawa iteration processes.

Theorem 8.6 *Let C be a nonempty, bounded, closed, and convex subset of a uniformly convex Banach space X. In addition, let us assume that the norm in X is monotone and has the monotone Opial property. Assume that $KIS(\mathcal{T}, \lambda, \{t_n\}, x_0)$ is the monotone Krasnoselskii-Ishikawa iteration process, where $\mathcal{T} \in m\mathcal{APNS}(C)$ is such that J is finitely generated by $G = \{\alpha_1, ..., \alpha_M\}$. Assume that T_s is continuous for each $s \in G$. Let $\{x_k\}$ be a sequence generated by this process. Assume that the sequence $\{t_k\}$ of elements from J has been chosen so that for every $s \in G$ there exists a strictly increasing, quasi-periodic sequence of natural numbers $\{j_k : k = 1, 2, \ldots\}$ such that $t_{j_{k+1}} = s + t_{j_k}$ for every $k \in \mathbb{N}$. Then, $x_k \rightharpoonup w$, where $w \in K$ is a common fixed point for $\mathcal{T}$ such that $x_0 \preceq w$. Moreover, w is a minimal common fixed point with this property.*

Proof It follows from Lemma 8.4 that $\lim_{k \to \infty} \|T_s(x_k) - x_k\| = 0$ for every element $s \in G$. Consider $y, z \in K$, two weak cluster points of $\{x_k\}$. Therefore, there exist two subsequences $\{y_k\}$ and $\{z_k\}$ of $\{x_k\}$ such that $y_k \rightharpoonup y$ and $z_k \rightharpoonup z$. By Theorem 8.5, $y, z \in F(\mathcal{T})$. By Lemma 8.2 the following limits exist:

$$r_1 = \lim_{k \to \infty} \|x_k - y\|, \quad r_2 = \lim_{k \to \infty} \|x_k - z\|.$$

We claim that $y = z$. Assume to the contrary that $y \neq z$. Observe that $x_k \preceq y$ and $x_k \preceq z$ for every k because $y, z \in K$. By the monotone Opial property we obtain the following:

$$r_1 = \liminf_{k \to \infty} \|y_k - y\| < \liminf_{k \to \infty} \|y_k - z\| = r_2$$
$$= \liminf_{k \to \infty} \|z_k - z\| < \liminf_{k \to \infty} \|z_k - y\| = r_1.$$

The contradiction implies $y = z$, which means that $\{x_k\}$ has at most one weak cluster point in K. Since K is a closed, bounded, convex subset of a uniformly convex Banach space, it is weakly compact. Hence, $\{x_k\}$ must have exactly one weak cluster point $w \in K$. This implies that $x_k \rightharpoonup w$. Applying Theorem 8.5 again, we conclude that $w \in F(\mathcal{T})$. It remains to prove that w is a minimal common fixed point greater than or equal to x_0. To this end, let us assume to the contrary that there exists $z \in F(\mathcal{T})$ such that $x_0 \preceq z$ and $z \prec w$. From Lemma 8.2, we infer that $x_k \preceq z$ for $k = 1, 2, \ldots$. By applying the monotone Opial property we obtain

$$\liminf_{n \to \infty} \|x_n - w\| < \liminf_{n \to \infty} \|x_n - z\|$$

On the other hand, since the norm is monotone, we conclude from $x_n \preceq z \prec w$ that

$$\|x_n - z\| \leq \|x_n - w\|.$$

The contradiction completes the proof. $\square$

Remark 8.4 Typical examples of Banach spaces for which our Theorem 8.6 can be applied are: l^p and $L^p[0, 1]$ for $p > 1$; Orlicz spaces L^φ with the Luxemburg norm if φ is strictly convex and satisfies the condition Δ_2 (see [24]), modular function spaces with the Δ_2 and $(UUC2)$ properties (see for example [54, 65]).

8.5 Strong-Convergence

Just as with the other iteration processes addressed in the previous chapters, assuming C is compact, we can establish a strong convergence result for the monotone Krasnoselskii-Ishikawa iteration process.

Theorem 8.7 *Let C be a nonempty, convex, and compact subset of a uniformly convex Banach space X. Assume that $KIS(\mathcal{T}, \lambda, \{t_n\}, x_0)$ is the monotone Krasnoselskii-Ishikawa iteration process, where $\mathcal{T} \in m\mathcal{APNS}(C)$ is such that J is finitely generated by $G = \{\alpha_1, ..., \alpha_M\}$. Assume that T_s is continuous for each $s \in G$. Let $\{x_k\}$ be a sequence generated by this process. Assume that the sequence $\{t_k\}$ of elements from J has been chosen so that for every $s \in G$ there exists a strictly increasing, quasi-periodic sequence of natural numbers $\{j_k : k = 1, 2, \ldots\}$ such that $t_{j_{k+1}} = s + t_{j_k}$ for every $k \in \mathbb{N}$. Then, there exists an element $x \in K$ which is a common fixed point for $\mathcal{T}$ such that $x_0 \preceq x$. Furthermore, the sequence $\{x_k\}$ strongly converges to x. If, in addition, the norm in X is monotone and has the monotone Opial property, then x is a minimal common fixed point with this property.*

Proof It follows from Lemma 8.4 that $\lim_{k\to\infty} \|T_s(x_k) - x_k\| = 0$ for every $s \in G$. Let us fix temporarily a number $s \in G$. By compactness of C, there exists a subsequence $\{x_{p_k}\}$ of $\{x_k\}$ and element $x \in C$ such that

$$\|T_s(x_{p_k}) - x\| \to 0 \text{ as } k \to \infty,$$

which, together with the fact that $\{x_k\}$ is an approximate fixed point sequence, implies that

$$\lim_{k\to\infty} \|x_{p_k} - x\| \leq \lim_{k\to\infty} \|x_{p_k} - T_s(x_{p_k})\| + \lim_{k\to\infty} \|T_s(x_{p_k}) - x\| = 0. \quad (8.31)$$

By applying the same reasoning as in Remark 8.3, we conclude that $x \in K$. This implies that $x_p \preceq x$ for every $p \in \mathbb{N}$, which means that all these elements are comparable. Consequently, utilising also the regularity of the sequence $\{x_k\}$, we have for any $t \in G$,

$$\|T_t(x) - x\| \leq \|T_t(x) - T_t(x_{p_k})\| + \|T_t(x_{p_k}) - x_{p_k}\| + \|x_{p_k} - x\|$$
$$\leq a_t(x)\|x_{p_k} - x\| + \|T_t(x_{p_k}) - x_{p_k}\| + \|x_{p_k} - x\|.$$

The right-hand side of this inequality tends to zero as $k \to \infty$, because $\|x_{p_k} - x\| \to 0$, and $\|T_t(x_{p_k}) - x_{p_k}\| \to 0$. Therefore, $T_t(x) = x$ for every $t \in G$. By (8.16), we conclude that $x \in F(\mathcal{T})$. It follows from Lemma 4.1 that $\lim_{k\to\infty} \|x_k - x\|$

exists. This, together with (8.31), implies that $\lim_{k\to\infty} \|x_k - x\| = 0$. The proof of the minimality of x, under the additional assumptions of monotonicity of the norm and the monotone Opial property, follows exactly the same reasoning as in the proof of Theorem 8.6. □

8.6 Notes

The research direction discussed in this chapter was pioneered by Ran and Reurings in [91], focusing on a class of matrix equations of the form:

$$Y = Q \pm \sum_{i=1}^{m} A_i F(Y) A_i,$$

where $Y \in H(n)$, the set of $n \times n$ Hermitian matrices, and $F : H(n) \to H(n)$ is a monotone function, meaning that $F(Y_1) \leq F(Y_2)$ if $Y_1 \leq Y_2$. This function maps the set of all $n \times n$ positive definite matrices $P(n)$ into itself, while $A_1, \ldots, A_m$ are arbitrary $n \times n$ matrices, and $Q \in P(n)$. The investigation of such matrix equations was initially motivated by various applications, including stochastic filtering, control theory, and dynamic programming [38]. Nieto and Rodriguez-Lopez [84] enhanced the results of Ran and Reurings, employing similar arguments to discover periodic solutions for a class of differential equations. Khamsi and Khan [52] applied this framework to establish the convergence of the Krasnoselskii-Ishikawa iteration process to fixed points of monotone nonexpansive mappings in L_1. Moreover, Bin Dehaish and Khamsi [30] demonstrated analogues of Browder and Goehde fixed point theorems for monotone nonexpansive mappings in uniformly convex hyperbolic spaces and uniformly convex Banach spaces in every direction. In relation to the linkage to the graph theory, see [2, 46]. The reader is also referred to the 2015 survey article by Bachar and Khamsi [8].

The foundational work on the fixed point theory for semigroups of monotone nonexpansive mappings was initiated by Bachar and Khamsi in their 2015 paper [7]. They introduced notable examples of such semigroups and established important results on approximate fixed point sequences. However, the iterative methods initially developed for single monotone mappings [30] proved inadequate for handling semigroups. This necessitated a specialised theory addressing common fixed point results for semigroups of monotone nonexpansive mappings. Such a theory was initiated by Kozlowski in [68], where the author proved the existence of common fixed points for monotone contractive and monotone nonexpansive semigroups of nonlinear mappings acting in uniformly convex Banach spaces. The results of [68] were extended by Bachar, Khamsi, Kozlowski and Bounkhel [9] to the case of uniformly convex in every direction Banach spaces. These fundamental existence results paved the way for the construction of such common fixed points. In [69], Kozlowski proposed a Krasnosel'skii-Ishikawa type algorithm for this construction and established its convergence in the sense of the weak topology in X. Further developments extended these results to modular and hyperbolic spaces [2, 4, 10].

8.6 Notes

As announced in the commentary preceding Theorem 8.3, the existence aspect of this result can be derived from the celebrated Knaster-Tarski Theorem. This fact was recognised in the 2018 paper by Espinola and Wiśnicki, which we quote below.

Proposition 8.2 ([37, Corollary 1]) *Let X be a Banach space with a partial order $\preceq$ and let τ be a Hausdorff topology on X such that all order intervals are τ-closed. Suppose C is a (nonempty) τ-compact subset of X and $\mathcal{F}$ a (nonempty) commutative family of monotone maps from C into C. Then $\mathcal{F}$ has a common fixed point if and only if there exists $c \in C$ such that $c \preceq f(c)$ for every $f \in \mathcal{F}$. Moreover, the set of common fixed points of $\mathcal{F}$ has a maximal element.*

Notably, the case of monotone pointwise Lipschitzian semigroups, as presented in this book, has not been published until now.

Chapter 9
Applications and Related Topics

Abstract In the first two sections of this chapter we focus on applying our theory to deterministic and stochastic differential equations and related dynamical systems. We consider the following Cauchy problem for an unknown function $u(x, \cdot) : [0, \infty) \to C$:

$$\begin{cases} u(x, 0) = x, \\ \dfrac{\partial u}{\partial t}(x, t) + (I - H)(u(x, t)) = 0, \end{cases}$$

where $H : C \to C$ is pointwise Lipschitzian mapping. We will discuss when the family of nonlinear mappings $\mathcal{T} = \{T_t : t \geq 0\}$, defined as $T_t(x) = u(x, t)$, forms a pointwise Lipschitzian semigroup, and in particular, when it becomes an asymptotic pointwise nonexpansive semigroup on C. In a seperate section, we address the question when the constructed semigroup is monotone pointwise Lipschitzian. In Sect. 9.3, we discuss how the theory, developed in this book, can be applied to the analysis of the long-term behaviour of semigroups associated with nonlinear Markov processes and related stochastic differential equations, such as the general kinetic equations in the weak form. The final section provides a high level reference to the parallel theories developed in metric, hyperbolic and modular function spaces.

9.1 Application to Differential Equations and Dynamical Systems

Let C denote, as usual, a nonempty, closed, bounded and convex subset of a Banach space X. In this section, we conjecture that a dynamical process can be defined by $T_t(x) = u(x, t)$ where $u(x, \cdot) : [0, \infty) \to C$ is a solution of the following initial value problem (IVP) for an unknown function $u(x, \cdot) : [0, \infty) \to C$:

$$\begin{cases} u(x, 0) = x, \\ \dfrac{\partial u}{\partial t}(x, t) + (I - H)(u(x, t)) = 0, \end{cases} \qquad (9.1)$$

© The Author(s), under exclusive license to Springer Nature Switzerland AG 2026
W. M. Kozlowski, *Fixed Points of Semigroups of Pointwise Lipschitzian Operators*,
SpringerBriefs in Mathematics, https://doi.org/10.1007/978-3-032-08869-7_9

where $H : C \to C$ is pointwise Lipschitzian mapping. Recall that this means that for every $x \in C$ there exists $1 \leq a(x)$ such that $\|H(x) - H(y)\| \leq a(x)\|x - y\|$ for all $y \in C$.

Our goal is to construct a solution $u(x, \cdot)$ for the (IVP) (9.1). To achieve this we need to construct $u(x, \cdot)$ such that

$$u(x, t) = e^{-t}x + \int_0^t e^{s-t} H(u(x, s)) ds. \qquad (9.2)$$

It is easy to demonstrate, employing the standard methods of the Bochner integration, that the above formula produces the required solution. As mentioned at the beginning of this section, we will discuss under what conditions, the family of nonlinear mappings $\mathcal{T} = \{T_t : t \geq 0\}$, defined as $T_t(x) = u(x, t)$, forms a pointwise Lipschitzian semigroup, and in particular, when it becomes an asymptotic pointwise nonexpansive semigroup on C.

Let us introduce the following convenient notations which will be used throughout this section. For any $t > 0$ we define the function

$$K(t) = 1 - e^{-t} = \int_0^t e^{s-t} ds. \qquad (9.3)$$

Let $A > 0$. For a Bochner measurable function $v : [0, A] \to X$, $t \in [0, A]$, and any $\tau = \{t_0, \ldots, t_n\}$, a subdivision of the interval $[0, t]$, we define the quantity

$$S_\tau(v)(t) = \sum_{i=0}^{n-1} (t_{i+1} - t_i) e^{t_i - t} v(t_i). \qquad (9.4)$$

We begin with the following technical lemma, which is essential for all the calculations in this section.

Lemma 9.1 *Let $f, g : [0, R] \to X$ be two Bochner-integrable $\|\cdot\|$-bounded functions. Then for every $t \in [0, R]$,*

$$\left\| e^{-t} g(t) + \int_0^t e^{s-t} f(s) ds \right\| \leq e^{-t} \|g(t)\| + K(t) \sup_{s \in [0, t]} \|f(s)\|. \qquad (9.5)$$

Proof Without any loss of generality we can assume that $\sup_{s \in [0,t]} \|f(s)\| < \infty$. Let $\tau = \{t_0, \ldots, t_n\}$ be a subdivision of the interval $[0, t]$. Define

$$Z_\tau(t) = e^{-t} g(t) + \sum_{i=0}^{n-1} (t_{i+1} - t_i) e^{t_i - t} f(t_i).$$

According to the Lebesgue Dominated Convergence Theorem for Bochner integrals, for every $t \in [0, R]$, $Z_\tau(t)$ converges in norm to $Z(t)$, where

9.1 Application to Differential Equations and Dynamical Systems

$$Z(t) = e^{-t}g(t) + \int_0^t e^{s-t} f(s)ds,$$

as $|\tau| = \sup_{0 \leq i \leq n} |t_{i+1} - t_i| \to 0$. Consequently,

$$\|Z(t)\| \leq \liminf_{|\tau| \to 0} \|Z_\tau(t)\| \leq e^{-t}\|g(t)\| + \liminf_{|\tau| \to 0}\left(\sum_{i=0}^{n-1}(t_{i+1} - t_i)e^{t_i - t}\right) \sup_{s \in [0,t]} \|f(s)\|$$

$$= e^{-t}\|g(t)\| + \int_0^t e^{s-t} ds \sup_{s \in [0,t]} \|f(s)\|,$$

thus proving (9.5), given the definition of $K(t)$ in (9.3). □

Throughout this section of the book, we will use the sequence of functions $u_n : C \times [0, R] \to C$, defined by the inductive formula,

$$\begin{cases} u_0(x, t) = x \\ u_{n+1}(x, t) = e^{-t}x + \int_0^t e^{s-t} H(u_n(x, s))ds. \end{cases} \quad (9.6)$$

In this context, let us define the following quantities:

$$M_n(t) = \sup_{s \in [0,t]} a(u_n(x, s)),$$

and

$$M(t) = \sup_{n \in \mathbb{N}_0} M_n(t).$$

Observe that for all n and t, both $M_n(t)$ and $M(t)$ are at least 1, and equal to 1 when H is nonexpansive.

Theorem 9.1 *Assume that*

$$K(t)M(t) < 1 \quad (9.7)$$

holds for every $t > 0$. Let $R > 0$ be fixed arbitrarily. Then, the sequence (9.6) converges uniformly in $t \in [0, R]$ to the solution $u(x, t)$, as given by the formula (9.2), of the initial value problem (9.1).

Proof Let $t \in [0, R]$. For a given subdivision τ of $[0, t]$, let us define functions

$$u_{n+1}^\tau(x, t) = e^{-t}x + S_\tau(H(u_n(x, \cdot)))(t).$$

By using convexity of C and recalling the definition of S_τ from (9.4), we can easily prove by induction that $u_n^\tau(x, t) \in C$. By the properties of the Bochner integral, we obtain

$$\|u_n^\tau(x, t) - u_n(x, t)\| \to 0,$$

as $|\tau| \to 0$. Since C is closed, it follows that $u_n(x,t) \in C$. We will prove now that the sequence $\{u_n(x,t)\}$ satisfies the Cauchy condition with respect to the norm in X. Since $K(R)M(R) < 1$, it suffices to validate the inequality

$$\|u_{n+p}(x,t) - u_n(x,t)\| \leq [K(R)M(R)]^n \text{diam}(C), \tag{9.8}$$

for any $p \in \mathbb{N}$ and all $t \in [0, R]$. We prove this by induction with respect to n with p fixed arbitrarily. For $n = 0$, the inequality (9.8) is trivially satisfied. Assume now (9.8) holds for $n = k$. Note that

$$u_{k+1+p}(x,t) - u_{k+1}(x,t) = \int_0^t e^{s-t}\Big(H(u_{k+p}(x,s)) - H(u_k(x,s))\Big)ds. \tag{9.9}$$

By applying (9.5) with $g(t) = 0$ and $f(t) = H(u_{k+p}(x,s)) - H(u_k(x,s))$, and using inequality (9.7) alongside our inductive assumption, we infer from (9.9) that

$$\|u_{k+1+p}(x,t) - u_{k+1}(x,t)\| \leq K(R) \sup_{s \in [0,R]} \|H(u_{k+p}(x,s)) \tag{9.10}$$

$$- H(u_k(x,s))\|$$
$$\leq K(R) \sup_{s \in [0,R]} a(u_{k+p}(x,s)) \sup_{s \in [0,R]} \|u_{k+p}(x,s)$$
$$- u_k(x,s)\|$$
$$\leq M(R)K(R)[K(R)M(R)]^k \text{diam}(C)$$
$$= [K(R)M(R)]^{k+1} \text{diam}(C).$$

This completes the proof of (9.8). Consequently, $\{u_n(x,t)\}$ is a Cauchy sequence. Since C is closed, it follows then that there exists an element $u(x,t) \in C$ such that $\|u_n(x,t) - u(x,t)\| \to 0$ as $n \to \infty$. Knowing this, take $p \to \infty$ in formula (9.8) to derive the estimation

$$\|u(x,t) - u_n(x,t)\| \leq K(R)^n M(R)^n \text{diam}(C). \tag{9.11}$$

From inequality (9.11) it follows that

$$\|H(u_n(x,s)) - H(u(x,s))\| \leq a(u_n(x,s))\|u_n(x,s) - u(x,s)\|$$
$$\leq M(R)^{n+1} K(R)^n \text{diam}(C),$$

which implies that both $\|u_n(x,s) - u(x,s)\|$ and $\|H(u_n(x,s)) - H(u(x,s))\|$ tend to zero uniformly in $t \in [0, R]$. Therefore, according to the Lebesgue Dominated Convergence Theorem, it follows from the definition of the sequence $\{u_n(x,t)\}$ in (9.6) that

$$u(x,t) = e^{-t}x + \int_0^t e^{s-t} H(u(x,s))ds,$$

9.1 Application to Differential Equations and Dynamical Systems

as claimed. Since $R > 0$ was chosen arbitrarily, it follows that the solution of the the Initial Value Problem (9.1) can be extended to a solution $u(x, t)$ defined on $[0, +\infty)$, such that its restriction to the interval $[0, R]$ is the uniform limit of the sequence $\{u_n(x, t)\}$. □

The next result demonstrates that, under certain conditions, each of the mappings $T_s(x) = u(x, t)$ is pointwise Lipschitzian.

Theorem 9.2 *Let us assume, in addition to (9.7), that $\prod_{i=0}^{\infty} M_i(t) < \infty$. Then, for every $s \geq 0$ there exists a finite number $A_s(x) \geq 1$ such that any $y \in C$,*

$$\|T_s(x) - T_s(y)\| \leq A_s(x)\|x - y\|.$$

Proof The following inequality is key for proving the assertion:

$$\|u_{n+1}(x, t) - u_{n+1}(y, t)\| \leq \prod_{i=0}^{n} M_i(t)\|x - y\|. \tag{9.12}$$

We will prove (9.12) by induction with respect to $n \in \mathbb{N}_0$. Let $n = 0$. By utilising the definition of the sequence $\{u_k(x, t)\}$ in (9.6) alongside the formula (9.5), we obtain

$$\|u_1(x, t) - u_1(y, t)\| \leq e^{-t}\|x - y\| + K(t) \sup_{s \in [0,t]} \|H(u_0(x, s)) - H(u_0(y, s))\|$$

$$\leq e^{-t}\|x - y\| + K(t) \sup_{s \in [0,t]} a(u_0(x, s))\|u_0(x, s) - u_0(y, s)\|$$

$$= e^{-t}\|x - y\| + K(t)M_0(t)\|x - y\|.$$

Assume that the inequality (9.12) holds for $n - 1$, where $n \geq 1$. We now aim to prove it for n. Following the same approach as for $n = 0$, and utilising the formula $K(t) = 1 - e^{-t}$, we can conclude that

$$\|u_{n+1}(x, t) - u_{n+1}(y, t)\| \leq e^{-t}\|x - y\| + K(t) \sup_{s \in [0,t]} \|H(u_n(x, s)) - H(u_n(y, s))\|$$

$$= e^{-t}\left(\|x - y\| - \sup_{s \in [0,t]} \|H(u_n(x, s)) - H(u_n(y, s))\|\right)$$

$$+ \sup_{s \in [0,t]} \|H(u_n(x, s)) - H(u_n(y, s))\|. \tag{9.13}$$

First, let us consider the case where

$$\sup_{s \in [0,t]} \|H(u_n(x, s)) - H(u_n(y, s))\| \leq \|x - y\|, \tag{9.14}$$

which by the top line of (9.13) implies that

$$\|u_{n+1}(x,t) - u_{n+1}(y,t)\| \le e^{-t}\|x-y\| + K(t)\|x-y\| = \|x-y\| \le \prod_{i=0}^{n} M_i(t)\|x-y\|,$$

since $M_i(t) \ge 1$. If the inequality (9.14) does not hold then, by using the bottom lines of (9.13) and the inductive assumption, we conclude that

$$\begin{aligned}
\|u_{n+1}(x,t) - u_{n+1}(y,t)\| &\le \sup_{s \in [0,t]} \|H(u_n(x,s)) - H(u_n(y,s))\| \\
&\le M_n(t) \sup_{s \in [0,t]} \|u_n(x,s) - u_n(y,s)\| \\
&\le M_n(t) \prod_{i=0}^{n-1} M_i(t)\|x-y\| = \prod_{i=0}^{n} M_i(t)\|x-y\|,
\end{aligned}$$

which completes the proof of (9.12). Taking the limit as $n \to \infty$ of both sides of this inequality we finally conclude that

$$\|T_t(x) - T_t(y)\| = \|u(x,t) - u(y,t)\| \le \prod_{i=0}^{\infty} M_i(t)\|x-y\| = A_t(x)\|x-y\|,$$

completing the proof. □

The next technical result will be of crucial importance for proving that the family of nonlinear mappings $\{T_t(x) = u(x,t)\}$ forms a pointwise Lipschitzian semigroup.

Lemma 9.2 *Under the assumptions of Theorem 9.1, the inequality*

$$\|u_n(u(x,\mu),t) - u_{n+m}(x,t+\mu)\|$$
$$\le \left(\sum_{i=n+1}^{n+m} K^i(\mu) M^i(\mu) \right) d + K^{n+1}(t) M^{n+1}(t) d,$$

holds for any $x \in C$, $t > 0$, $\mu > 0$, $n \in \mathbb{N}$, $m \in \mathbb{N}$, *where* $d = \text{diam}(C)$.

Proof The proof is by induction on $n \in \mathbb{N}_0$. Assume first that $n = 0$. Use the recurrent definition of the sequence $\{u_m\}$ to obtain the formula

$$u_0(u(x,\mu),t) - u_m(x,t+\mu) = u(x,\mu) - u_m(x,t+\mu) \quad (9.15)$$
$$= u(x,\mu) - e^{-t-\mu}x - \int_0^{t+\mu} e^{s-t-\mu} H(u_{m-1}(x,s)) ds.$$

By applying the change of variable formula, we have

$$\int_0^{t+\mu} e^{s-t-\mu} H(u_{m-1}(x,s)) ds = e^{-t} \int_0^{\mu} e^{s-\mu} H(u_{m-1}(x,s)) ds \quad (9.16)$$
$$+ e^{-t} \int_0^{t} e^s H(u_{m-1}(x,s+\mu)) ds.$$

9.1 Application to Differential Equations and Dynamical Systems

Based on the definition of u_m and utilising (9.16), we conclude that the following calculation is valid

$$u_m(x, t+\mu) = e^{-t}\left(e^{-\mu}x + \int_0^\mu e^{s-\mu} H(u_{m-1}(x,s))ds\right) \qquad (9.17)$$

$$+ e^{-t}\int_0^t e^s H(u_{m-1}(x, s+\mu))ds$$

$$= e^{-t}u_m(x,\mu) + e^{-t}\int_0^t e^s H(u_{m-1}(x, s+\mu))ds.$$

Substitute (9.17) into (9.15) to obtain

$$u_0(u(x,\mu), t) - u_m(x, t+\mu) \qquad (9.18)$$

$$= u(x,\mu) - e^{-t}u_m(x,\mu) - e^{-t}\int_0^t e^s H(u_{m-1}(x, s+\mu))ds$$

$$= e^{-t}\Big(u(x,\mu) - u_m(x,\mu)\Big) + \int_0^t e^{s-t}\Big(u(x,\mu) - H(u_{m-1}(x, s+\mu))\Big)ds,$$

where we used the identity

$$u(x,\mu) = e^{-t}u(x,\mu) + \int_0^t e^{s-t} u(x,\mu) ds.$$

Apply Lemma 9.1 with

$$f(t) = u(x,\mu) - H(u_{m-1}(x, t+\mu)),$$
$$g(t) = u(x,\mu) - u_m(x,\mu),$$

to obtain the following series of inequalities. Observe how we use (9.5) and (9.18) in the following calculations,

$$\|u_0(u(x,\mu), t) - u_m(x, t+\mu)\| \qquad (9.19)$$

$$= \left\| e^{-t} g(t) + \int_0^t e^{s-t} f(s) ds \right\|$$

$$\le e^{-t}\|g(t)\| + K(t) \sup_{s\in[0,t]} \|f(s)\|$$

$$= e^{-t}\|u(x,\mu) - u_m(x,\mu)\| + K(t) \sup_{s\in[0,t]} \|u(x,\mu) - H(u_{m-1}(x, s+\mu))\|$$

$$\le e^{-t}\|u(x,\mu) - u_m(x,\mu)\| + K(t)d.$$

By applying inequality (9.11) to the right-hand side of (9.19) and considering that $M(t) \ge 1$, we conclude that

$$\|u_0(u(x,\mu),t) - u_m(x,t+\mu)\| \le K^m(\mu) M^m(\mu) d + K(t) d$$
$$\le \Big(\sum_{i=1}^{m} K^\mu(t) M^i(\mu)\Big) d + K(t) M(t) d,$$

which gives us the desired inequality (9.15) for $n = 0$. Assume now that (9.15) holds for $n \in \mathbb{N}$ and let us prove it for $n+1$. By using the definition of the recurrent sequence together with the formula (9.17) we obtain

$$u_{n+1}(u(x,\mu),t) - u_{n+m+1}(x,t+\mu) \qquad (9.20)$$
$$= e^{-t} u(x,\mu) + \int_0^t e^{s-t} H(u_n(u(x,\mu),s)) ds - u_{n+m+1}(x,t+\mu)$$
$$= e^{-t} u(x,\mu) + \int_0^t e^{s-t} H(u_n(u(x,\mu),s)) ds$$
$$\quad - \Big(e^{-t-\mu} u_{n+m+1}(x,\mu) + e^{-t} \int_0^t e^s H(u_{n+m}(x,s+\mu)) ds \Big)$$
$$= e^{-t}(u(x,\mu) - u_{n+m+1}(x,\mu)) + \int_0^t e^{s-t}[H(u_n(u(x,\mu),s)) - H(u_{n+m}(x,s+\mu)))] ds.$$

By applying Lemma 9.1 with

$$f(t) = H(u_n(u(x,\mu),t)) - H(u_{n+m}(x,t+\mu)),$$
$$g(t) = u(x,\mu) - u_{n+m+1}(x,\mu),$$

we conclude from (9.20) that

$$\|u_{n+1}(u(x,\mu),t) - u_{n+m+1}(x,t+\mu)\| \qquad (9.21)$$
$$\le e^{-t} \|u(x,\mu) - u_{n+m+1}(x,\mu)\| + K(t) \sup_{s\in[0,t]} \|H(u_n(u(x,\mu),s)) - H(u_{n+m}(x,s+\mu))\|$$
$$\le e^{-t} \|u(x,\mu) - u_{n+m+1}(x,\mu)\| + K(t) \sup_{s\in[0,t]} a(u_n(u(x,\mu),s)) \|u_n(u(x,\mu),s) - u_{n+m}(x,s+\mu)\|$$
$$\le e^{-t} K(\mu)^{n+m+1} M(\mu)^{n+m+1} d + K(t) M(t) \sup_{s\in[0,t]} \|u_n(u(x,\mu),s) - u_{n+m}(x,s+\mu)\|$$
$$\le e^{-t} K(\mu)^{n+m+1} M(\mu)^{n+m+1} d + K(t) M(t) \Big(d \sum_{i=n+1}^{n+m} K^i(\mu) M^i(\mu) + K^{n+1}(t) M^{n+1}(t) d \Big)$$
$$\le \Big(\sum_{i=n+1}^{n+m+1} K^i(\mu) M^i(\mu) \Big) d + K^{n+2}(t) M^{n+2}(t) d,$$

where we also used (9.11) and the inductive assumption. The proof is complete. $\square$

Theorem 9.3 *Denote $T_t(x) = u(x,t)$, where $t \ge 0$, $x \in C$ and $u(x,t)$ is a solution of the initial value problem (9.1) given by the formula (9.2). Then, under the assumptions of Theorem 9.2, the family of mappings $\{T_t\}_{t\ge 0}$ forms a pointwise Lipschitzian semigroup.*

Proof We have already established in Theorem 9.2 that each T_t is a pointwise Lipschitzian mapping. Clearly $T_0(x) = u(x, 0) = x$. The function $t \mapsto T_t(x) = u(x, t)$ is norm continuous as the almost uniform limit of continuous functions $t \mapsto u_n(x, t)$. It remains to prove that $T_{t+s}(x) = T_t(T_s(x))$. Let us fix temporarily $n \in \mathbb{N}$. By using (9.15) from Lemma 9.2 we have

$$\begin{aligned} \|u_n(u(x, \mu), t) - T_{t+\mu}(x)\| &= \|u_n(u(x, \mu), t) - u(x, t + \mu)\| \\ &= \lim_{m \to \infty} \|u_n(u(x, \mu), t) - u_{n+m}(x, t + \mu)\| \\ &\leq \lim_{m \to \infty} \left(\sum_{i=n+1}^{n+m} K^i(\mu) M^i(\mu) \right) d + K^{n+1}(t) M^{n+1}(t) d \to 0, \end{aligned}$$

as $n \to \infty$ because $K(s)M(s) < 1$ for any $s > 0$. On the other hand, it follows from (9.11) that

$$\begin{aligned} \|u_n(u(x, \mu), t) - T_t(T_\mu(x))\| &= \|u_n(T_\mu(x), t) - T_t(T_\mu(x))\| \\ &= \|u_n(T_\mu(x), t) - u(T_\mu(x), t)\| \to 0. \end{aligned}$$

From the uniqueness of the limit it follows that $T_t(T_s(x)) = T_{t+s}(x)$, as claimed. In addition, note that $\{T_t\}_{t \geq 0}$ becomes asymptotic pointwise nonexpansive when $A_t(x) = \prod_{i=0}^{\infty} M_i(t) \to 1$ as $t \to \infty$. □

Example 9.1 Let us shed a bit more light on the above theorem by providing some concrete examples for the constants $M_n(t)$. Suppose that they can be estimated by

$$M_n(t) \leq \left(1 + \frac{e^{-t}}{n^2 + 1}\right).$$

It is easy to see that $M(t) \leq (1 + e^{-t})$, which implies that $M(t)K(t) \leq 1 - e^{-2t} < 1$. To ensure we can use Theorem 9.3, we need to estimate $A_t(x) = \prod_{n=0}^{\infty} M_n(t)$, which is done below,

$$\begin{aligned} A_t(x) = \prod_{n=0}^{\infty} M_n(t) &\leq \prod_{n=0}^{\infty} \left(1 + \frac{e^{-t}}{n^2 + 1}\right) \\ &\leq \exp\left[e^{-t}\left(1 + \sum_{n=1}^{\infty} \frac{1}{n^2}\right)\right] = \exp\left[e^{-t}\left(1 + \frac{\pi^2}{6}\right)\right] < \infty. \end{aligned}$$

Note that each $A_t(x) \geq 1$ and that $A_t(x) \to 1$ as $t \to +\infty$. It follows from Theorem 9.2 that in this case the semigroup generated by this system would be asymptotic pointwise nonexpansive.

Let us summarise the results obtained in this section. First, we established that the initial value problem (IVP) defined in (9.1) has the solution $u(x, t)$. Next, we showed that, under certain conditions, the solution set $T_t(x) = u(x, t)$ forms a pointwise

Lipschitz semigroup, providing a non-trivial and significant example for applications in differential equations and dynamic systems. We observed that, in some cases, this semigroup may be asymptotically pointwise nonexpansive. If this holds, we can apply the theory presented in this book to demonstrate existence and construct a common fixed point for $T_t(x)$. Recall that such common fixed points correspond to the stationary points of the dynamic system evolution represented by the function $t \mapsto T_t(x)$ with an initial point at x.

Questions regarding the construction, generation, and approximation of semigroups of nonlinear operators in relation to differential equations have been explored for many decades, beginning with the pioneering works of Kato [47], Barbu [11], and Nisio [85]. This area has seen numerous later studies, including the influential book by Reich and Shoikhet [93], as well as recent investigations into Lipschitzian (but not pointwise Lipschitzian) semigroups [15, 16, 34]. The approach presented in this section traces its roots to studies on nonexpansive semigroups by Crandall and Pazy [26] and Reich [92]. The methods employed here build on techniques developed for the nonexpansive case in the works of Khamsi [51], Bachar and Khamsi [6], and Kozlowski [63, 66, 68]. Notably, the case of pointwise Lipschitzian semigroups, as discussed in this section, has not previously been studied.

9.2 The Monotone Case

In this section we assume that $(X, \preceq)$ is an ordered Banach space, as defined below.

Definition 9.1 We say that $(X, \preceq)$ is an ordered Banach space if the following two conditions are satisfied:

1. $x \preceq y \Rightarrow \alpha x \preceq \alpha y$ for $\alpha \geq 0$,
2. $a \preceq b,\ c \preceq d \Rightarrow a + c \preceq b + d$.

Note that this is a very typical situation; consider for example the usual function and sequence spaces like L^p or l^p with the pointwise order. Observe that the condition (8.1) of Sect. 8.2 follows from the above assumptions, implying that all order intervals are convex. As usual, we assume that all order intervals are closed.

In addition to the assumptions of Theorem 9.3, we assume that H is monotone pointwise Lipschitzian mapping in the sense of Definition 8.2. Consequently, we assume that all elements $x \in C$ satisfy $x \preceq H(x)$. Our goal is to demonstrate that the semigroup generated by the solution to $u(x, t)$ of the initial value problem (9.1) given by the formula (9.2) forms a monotone pointwise Lipschitzian semigroup defined by $T_t(x) = u(x, t)$, where $x \in C$. Given that the general reasoning aligns with the flow established in the previous section, we shall refer to the results presented therein. However, we will elaborate in detail on the points concerning monotonicity. The section will conclude with an example that illustrates the significance of the monotone case.

Let us begin with a series of supporting results.

9.2 The Monotone Case

Proposition 9.1 *Let $\{u_n\}$ be the sequence defined by (9.6). For every $n \in \mathbb{N}_0$ and every $t \in [0, R]$,*

$$u_n(x, t) \preceq u_{n+1}(x, t). \tag{9.22}$$

Proof We will prove (9.22) by induction. It follows from $x \preceq H(x)$ that

$$u_0(x, t) = x = e^{-t}x + (1 - e^{-t})x \preceq e^{-t}x + \int_0^t e^{s-t} H(x) ds = u_1(x, t).$$

Assume now that (9.22) holds for $n = k$. Since H is monotone, it follows that

$$u_{k+1}(x, t) = e^{-t}x + \int_0^t e^{s-t} H(u_k(x, s)) ds$$

$$\preceq e^{-t}x + \int_0^t e^{s-t} H(u_{k+1}(x, s)) ds = u_{k+2}(x, t). \qquad \square$$

Remark 9.1 Note that (9.22) can be rewritten as

$$x = u_0(x, t) \preceq u_1(x, t) \preceq \ldots u_n(x, t) \preceq u_{n+1}(x, t) \preceq \ldots,$$

which implies that x and all $u_n(x, t)$ are comparable to each other (in a monotone way) and that they belong to the order interval $[x, \rightarrow)$. Since $u_n(x, t) \in [x, \rightarrow)$ and $[x, \rightarrow)$ is closed it follows that $x \preceq u(x, t)$, where $u(x, t)$ is an uniform limit of $\{u_n(x, t)\}$ on $[0, R]$, existing by Theorem 9.1. Note that we required the monotonicity of $u_n(x, t)$, as established in Proposition 9.1, to ensure the validity of the inequality (9.10) in the monotone case. Once this condition is satisfied, Theorem 9.1 guarantees that $u(x, t)$, being the uniform limit of $u_n(x, t)$, is a solution to our initial value problem.

Proposition 9.2 *If $y \preceq x$ then*

$$u_n(y, t) \preceq u_n(x, t) \tag{9.23}$$

holds for every $n \in \mathbb{N}_0$ and every $t \in [0, R]$.

Proof We will proceed by induction with respect to $n \in \mathbb{N}_0$. Since $u_0(y, t) = y \preceq x = u_0(x, t)$, it follows that (9.23) holds for $n = 0$. Assume now it is valid for $n \in \mathbb{N}_0$. The following inequality holds in view of the definition of the sequence $\{u_n\}$, properties of the order, monotonicity of H, and the inductive assumption,

$$u_{n+1}(y, t) = e^{-t}y + \int_0^t e^{s-t} H(u_n(y, s)) ds \preceq e^{-t}x + \int_0^t e^{s-t} H(u_n(x, s)) ds = u_{n+1}(x, t),$$

finishing the proof. $\qquad \square$

Proposition 9.3 *The inequality*

$$u_n(x, t + \mu) \preceq u_n(u(x, \mu), t) \tag{9.24}$$

holds for every $n \in \mathbb{N}_0$ and all $t, \mu \in [0, R]$ such that $t + \mu \leq R$.

Proof The formula (9.24) holds for $n = 0$ because $u_0(x, t + \mu) = x \preceq u(x, \mu) = u_0(u(x, \mu), t)$, which was established in Remark 9.1. To prove our assertion by induction, let us suppose that it is true for $n \in \mathbb{N}_0$. By utilising the inductive assumption and the calculations already proved and used in this chapter (see the proof of Lemma 9.2), we obtain the following sequence of inequalities,

$$u_{n+1}(x, t + \mu) = e^{-t} u_{n+1}(x, \mu) + \int_0^t e^{s-t} H(u_n(x, s + \mu)) ds$$

$$\preceq e^{-t} u_{n+1}(x, \mu) + \int_0^t e^{s-t} H(u_n(u(x, \mu), s)) ds$$

$$\preceq e^{-t} u_{n+1}(x, \mu) + \left[\int_0^t e^{s-t} H(u_n(u(x, \mu), s)) ds + e^{-t} u(x, \mu) \right] - e^{-t} u(x, \mu)$$

$$= e^{-t} u_{n+1}(x, \mu) + u_{n+1}(u(x, \mu)) - e^{-t} u(x, \mu)$$

$$\preceq u_{n+1}(u(x, \mu)) + e^{-t}(u_{n+1}(x, \mu) - u(x, \mu))$$

$$\preceq u_{n+1}(u(x, \mu), t),$$

where we also used the inequality $u_{n+1}(x, \mu) \preceq u(x, \mu)$. The proof is complete. □

Proposition 9.4 *The inequality*

$$u_{n+m}(x, t + \mu) \preceq u_n(u(x, \mu), t). \tag{9.25}$$

holds for every $n, m \in \mathbb{N}_0$ and all $t, \mu \in [0, R]$ such that $t + \mu \leq R$.

Proof Fix $n \in \mathbb{N}_0$. The proof is by induction with respect to $m \in \mathbb{N}_0$. For $m = 0$, the inequality (9.25) follows from Proposition 9.3. The proof of the inductive step follows the pattern used in the proof of Proposition 9.3, with u_n replaced now by u_{n+m}. □

Theorem 9.4 *Under the assumptions used in this section, $u(x, t)$ defined by the formula (9.2) is a solution to the initial value problem (9.1). Moreover, $u(x, t)$ satisfies $x \preceq u(x, t)$ for every $t \geq 0$. Furthermore, the family of mappings $\{T_t\}_{t \geq 0}$, generated by this solution, forms a monotone pointwise Lipschitzian semigroup.*

Proof The proof employs a similar line of reasoning as used in Theorem 9.3. However, it is important to proceed carefully to ensure that all monotonicity conditions are satisfied at each step. Lemma 9.1 does not involve any monotonicity related conditions. Theorem 9.1 and Theorem 9.2 still hold since the pointwise Lipschitzian conditions are only applied to comparable elements of $\{u_k(x, s)\}$ and $u(x, s)$. In the proof of Lemma 9.2, we need to apply Proposition 9.4 to make sure that the inequality (9.21) remains valid. As discussed in Remark 9.1, Theorem 9.1 guarantees that $u(x, t)$, being the uniform limit of the nondecreasing sequence $u_n(x, t)$, is a solution to our initial value problem (see also Proposition 9.1). The inequality $x \preceq u(x, t)$ is considered in Remark 9.1. The proofs of the semigroup conditions $T_0(x) = x$ and $T_{t+s}(x) = T_t(T_s(x))$ remain unchanged. It remains to be proved that each T_t is a monotone pointwise Lipschitzian mapping. To prove that T_t is a monotone mapping, observe that, by Proposition 9.2, if $x \preceq y$ then $u_n(x, t) \preceq u_n(y, t)$

9.2 The Monotone Case

for every $n \in \mathbb{N}_0$. Both sequences are nondecreasing, and convergent to $u(y, t)$ and $u(x, t)$, respectively. Given that all order intervals are closed, it is evident that in this context, $T_t(x) = u(x, t) \preceq u(y, t) = T_t(y)$. By examining the proof of Theorem 9.2, we deduce that for every $y \preceq x$, there exists a function $a(x)$ such that $\|T_t(x) - T_t(y)\| \leq a(x)\|x - y\|$. Thus, T_t is confirmed to be a monotone pointwise Lipschitzian mapping, as asserted. This completes the proof. □

The following example, adapted from the work by Bachar and Khamsi [6], sheds more light on the benefits of working with monotone pointwise Lipschitzian mappings and semigroups of such mappings.

Example 9.2 Consider a measurable function $f : [0, 1] \times [0, 1] \times \mathbb{R} \to [0, \frac{1}{2}]$ such that for all $t, s \in [0, 1]$ and all $v \leq u$,

$$f(t, s, u) - f(t, s, v) \geq 0. \tag{9.26}$$

Assume that for every u there exists $\gamma(u) \geq 0$ such that

$$f(t, s, u) - f(t, s, v) \leq \gamma(u)(u - v)$$

holds for every $v \leq u$. Let $g : [0, 1] \to [0, \frac{1}{2}]$ be measurable. Define

$$C = \{x \in L^2([0, 1]) : 0 \leq x(t) \leq 1 \text{ a.e.}\}.$$

Note that C is convex and bounded as a closed interval in $L^2([0, 1])$, where the order is defined by $x \preceq y$ if $x(t) \leq y(t)$ almost everywhere; recall Remark 8.1. Define a mapping $H : C \to C$ by

$$H(x)(t) = g(t) + \int_0^1 f(t, s, x(s)) ds. \tag{9.27}$$

It follows from (9.26) that H is a monotone operator. We will show that, under some natural conditions, H becomes a monotone pointwise Lipschitzian mapping. Assume that $y \preceq x$, where $y, x \in C$. By using the Cauchy-Schwarz inequality, we see that

$$\|H(x) - H(y)\|_{L^2}^2 = \int_0^1 \left(\int_0^1 f(t, s, x(s)) - f(t, s, y(s)) ds \right)^2 dt$$
$$\leq \left(\int_0^1 \gamma(x(s))(x(s) - y(s)) ds \right)^2 \leq \left(\int_0^1 \gamma^2(x(s)) ds \right) \|x - y\|_{L^2}^2.$$

Therefore, H is a monotone pointwise Lipschitzian mapping provided that

$$a(x) = \left(\int_0^1 \gamma^2(x(s)) ds \right)^{\frac{1}{2}} < \infty, \tag{9.28}$$

for every $x \in C$. Note that this is for example the case if the function $\gamma : [0, \frac{1}{2}] \to [0, \infty)$ is a bounded function. As a simple, concrete example of a function f satisfying the above condition we can take

$$f(t, s, u) = h(t, s) + Mu^2, \tag{9.29}$$

where $M \geq 1$ and $h \in L^2([0, 1] \times [0, 1])$. Clearly, f satisfies (9.26). Furthermore, for $u \geq v$ we have

$$0 \leq f(t, s, u) - f(t, s, v) = M(u^2 - v^2) = M(u + v)(u - v) \leq 2Mu(u - v).$$

We can take, therefore $\gamma(u) = 2Mu$. It follows from (9.28) that $a(x) = 2M\|x\|_{L^2}$. This implies that the mapping H defined by (9.27) with the function $f(t, s, u)$ defined in (9.29) is a monotone pointwise Lipschitzian mapping.

Examples like these illustrate how our theory, developed in Chap. 8, can be utilised for solving integral equations of the form

$$x(t) = g(t) + \int_0^1 f(t, s, x(s)) ds.$$

Our fixed point existence results from Sect. 8.3 offer further insights into these fixed points. Recall that a fixed point z satisfies the inequality $x \preceq z$, where x is a known element with the condition $x \preceq T_t(x)$. This additional information is particularly valuable in applications, such as when searching for a non-negative solution. Our Theorem 9.4 demonstrates that monotone pointwise Lipschitzian mappings, such as those presented in Example 9.2, can—under certain conditions—produce monotone pointwise Lipschitzian semigroups. As previously noted, focusing on monotone pointwise Lipschitzian semigroups is advantageous because the pointwise Lipschitzian condition must only be verified on a significantly smaller set associated with the initial element x. Similarly to the situation in Chap. 8, and indeed throughout the entire book, we should recognise that all the results in this section apply to the simpler case of monotone nonexpansive semigroups, where $a(x) = 1$. In this scenario, it is clear that most of the assumptions of Theorem 9.4 are trivially satisfied.

The results presented in this section are novel. The application of monotone nonexpansive semigroups in Banach spaces to differential equations and dynamical systems was explored by Kozlowski in his 2018 paper [68]. The findings in the current section can be viewed as an extension of those in [68].

9.3 Semigroups Corresponding to Stochastic Processes

In this section, we will outline possible applications of the results discussed in this book to considerations related to the long-term behaviour of semigroups associated

9.3 Semigroups Corresponding to Stochastic Processes

with nonlinear stochastic processes. A classical linear Markov process is a stochastic process where the future states only depend on the current state and not on the past states. Nonlinear Markov processes are Markov processes where the future depends not only on its present state, but also on its present distribution. This type of Markov process was initially introduced by McKean in [83] in the context of mechanical transport problems and associated nonlinear parabolic equations. Nonlinear Markov processes have been the subject of intensive study over recent decades. For further information, please refer to the monograph by Kolokoltsov [56], as well as more recent works by Kolokoltsov [57], Carmona and Delarue [23], Pham and Wei [90], and Neumann [86], along with the literature cited therein. Nonlinear Markov chains and related Markov semigroups arise naturally in several branches of applied sciences, for example, in evolutionary biology, epidemiology, and game theory.

A nonlinear Markov chain is characterised by a continuous family of nonlinear transition probabilities which form a nonlinear Markov semigroup, provided the relevant Chapman-Kolmogorov equation is satisfied. As an important example, one can consider the general kinetic equation in the weak equation form

$$\frac{d}{dt}(f, \mu_t) = (L_\mu f, \mu_t), \quad \mu_t \in \mathcal{P}(\mathbb{R}^d), \quad \mu_0 = \mu, \quad (9.30)$$

where

$$L_\mu f(x) = \frac{1}{2}(G(x, \mu)\nabla, \nabla)f(x) + (b(x, \mu), \nabla f(x)) \\ + \int (f(x + y) - f(x) - (\nabla f(x), y)1_{B_1}(y))\nu(x, \mu, dy),$$

function $f \in C_c^2(\mathbb{R}^d)$, $G(x)$ is a symmetric non-negative matrix and $\nu(x, \cdot)$ is a Borel measure on $\mathbb{R}^d$ (called Lévy measure). Furthermore, μ stands for a finite positive measure on $\mathbb{R}^d$, the pairing (f, μ) denotes the usual integration, B_1 is the unit ball in $\mathbb{R}^d$ and 1_{B_1} is its indicator function. Using some well-posedness assumptions, Kolokoltsov proved in [57, Theorem 1.2] that the resolving operators $U_t : \mu \mapsto \mu_t$ of the Cauchy problem (9.30) form a nonlinear Markov semigroup. These equations and the associated nonlinear Markov semigroups play a crucial role in the theory of interacting particles, encompassing all cases of positivity-preserving evolution on measures, provided certain mild regularity assumptions are met. Notable examples include prominent models based on the Boltzmann, Vlasov, Smoluchowski, Landau-Fokker-Planck, and McKean diffusion equations.

To explore how these results relate to the theory presented in this book, we turn our attention to a simpler yet still significant case of nonlinear Markov chains in continuous time with a finite state space. In our discussion, we will reference the terminology and results of the 2023 paper by Neumann [86]. Let $\mathcal{S} = \{1, ..., S\}$ be the state space of the nonlinear Markov chain. By $\mathcal{P}(\mathcal{S})$ let us denote the probability simplex over this state space. Such a nonlinear Markov chain is characterised by a continuous family of nonlinear transition probabilities $P(t, m) = \{P_{ij}(t, m) : i, j \in \mathcal{S}\}$, which is a family of stochastic matrices. Assume that $P(t,m)$ depends continuously on $t \geq 0$ and $m \in \mathcal{P}(S)$, and that it satisfies the nonlinear Chapman-Kolmogorov equation,

$$\sum_{i \in S} m_i P_{ij}(t+s, m) = \sum_{i,k \in S} m_i P_{ik}(t, m) P_{kj}\left(s, \sum_{l \in S} m_l P_l(t, m)\right). \tag{9.31}$$

Transitional probability $P_{ij}(t,m)$ is, as usual, interpreted as the probability that the process is in state j at time t given that the initial state was i and the initial distribution of the process was m. We recognise immediately that the nonlinear Chapman-Kolmogorov equation (9.31) implies the semigroup property. Indeed, it is not difficult to prove that the following formula:

$$(T_t(m))_{j \in S} = \sum_{i \in S} m_i P_{ij}(t, m) \tag{9.32}$$

defines a nonlinear Markov semigroup of continuous mappings $\mathcal{T} = \{T_t\}_{t \geq 0}$ acting within $\mathcal{P}(S)$. To examine the long-term-or using our terminology—asymptotic behaviour of this nonlinear Markov process, the author of [86] considers two characteristics of the semigroup $\mathcal{T}$. The first one relates to the question of existence and uniqueness, with respect to the initial marginal distribution m_0 at time $t_0 = 0$, of the stationary points of the process, which are the common fixed points of $\mathcal{T}$. A stationary point of this nature is referred to, in probabilistic terminology, as the invariant distribution. The paper [86] presents an example where, in contrast to the linear case, the marginal distributions do not converge but are instead periodic. In our terminology, this means that there exist periodic points of $\mathcal{T}$ that are not common fixed points of this semigroup. This is not surprising, as we know from Theorem 2.1 that only the asymptotic pointwise contractiveness of $\mathcal{T}$ can guarantee the uniqueness of a common fixed point. As discussed in Chap. 2, periodic points occur when $T_{t_1}(x) = x$ for some $t_1 \geq 0$, but not for all $t \geq 0$. In this case, the period is equal to t_1. There is nothing to suggest that, in general, the semigroup $\mathcal{T}$ defined in (9.32) must be asymptotic pointwise contractive. The second peculiarity noted by this paper pertains to the probabilistic notion of strong ergodicity. In our terminology, for the nonlinear Markov process to be strongly ergodic, we need existence and uniqueness (with respect to m) of the strong limit of the Picard's orbits $T_t(m)$. As discussed in Sect. 4.1, this generally does not hold, except when $\mathcal{T}$ is an asymptotic pointwise contractive semigroup. Thus, it is not surprising that the paper [86] provides examples of nonlinear Markov processes, even with very small state spaces, that exhibit different trajectories and limiting points for different initial conditions.

As appreciated in the cited studies, it is the nonlinear character of the considered processes that is at the core of the observed long-term probabilistic peculiarities. Therefore, it is quite natural that the asymptotic approach to semigroups of nonlinear mappings, as presented in this book, can provide a suitable toolset for dealing with such phenomena.

While further research in this area is needed, it is evident from the discussion above that combining probabilistic and analytic approaches can yield deeper insights and lead to more informative and less restrictive results. Specifically, applying the general semigroup fixed point results, presented in this book, can clarify when we can expect invariant distributions and strong ergodicity to hold, not only in the case of finite state models but also in more general infinite state space models, such as those defined by the general kinetic equations discussed at the beginning of this section.

9.4 Beyond Banach Spaces

For the sake of simplicity and consistency, this book primarily operates within the framework of Banach spaces. However, the approach presented here has also been applied in other types of spaces. While it is impossible to thoroughly address all these results in such a brief note, some significant threads of research are worth mentioning. The reader may find it beneficial to explore these trends through further study and research activities.

In the context of metric spaces, relevant research originated in [44] and was subsequently developed further in [29, 72] and extended to various types of hyperbolic spaces in several works including [10, 97]; in this context, see also the influential books [41, 42].

Significant advancements have been made in the context of modular function spaces, beginning with the book [58] and the seminal papers [51, 55]. For a comprehensive overview, please refer to the extensive list of references in [54]. We also encourage the reader to explore additional papers that address this subject matter, including [5, 31, 32, 60, 63, 65, 66, 70, 71, 73, 75], along with the literature cited within those works.

It is worth mentioning that the book [54] discusses applications to differential equations and dynamical systems generated by nonexpansive (in the modular sense) mappings, and elaborates on dynamical processes defined by nonlinear integral Urysohn operators. These results are based on earlier works by Khamsi and Kozlowski, as documented in [51, 63, 66]. Note also that several papers, including [4, 6, 52], address the case of monotone operators acting within modular function spaces. Advantages of using modular function spaces include their flexibility. For example, the Urysohn operator, mentioned above, defines itself a modular space in which this operator has properties allowing application of the general results of the common fixed point theory.

References

1. Aizenberg, L., Reich, S., Shoikhet, D.: One-sided estimates for the existence of null points of holomorphic mappings in Banach spaces. J. Math. Anal. Appl., **203**, 38–54 (1996)
2. Alfuraidan, M.R., and Khamsi, M.A.: Fixed points of monotone nonexpansive mappings on a hyperbolic metric space with a graph. Fixed Point Theory Appl., **2015:44**, (2015)
3. Alfuraidan, M.R., Khamsi, M.A.: Fibonacci-Mann iteration for monotone asymptotically nonexpansive mappings. Bull. Austral. Math. Soc., **96.2**, 307–316 (2017)
4. Alfuraidan, M.R., Khamsi, M.A., Kozlowski, W.M.: On monotone mappings in modular function spaces. In: Advances in Metric Fixed Point Theory and Applications. Springer, Singapore, 217–240 (2021)
5. Alsulami, S.M., Kozlowski, W.M.: On the set of common fixed points of semigroups of nonlinear mappings in modular function spaces. Fixed Point Theory Appl., **2014:4**, (2014)
6. Bachar, M., Khamsi, M.A.: Fixed Points of Monotone Mappings and Application to Integral Equations. Fixed Point Theory Applications, **2015:110**, (2015)
7. Bachar, M., Khamsi, M.A.: On common approximate fixed points of monotone nonexpansive semigroups in Banach spaces. Fixed Point Theory Appl., **2015:160**, (2015)
8. Bachar, M., Khamsi, M.A.: Recent contributions to fixed point theory of monotone mappings. J. Fixed Point Theory Appl., **19.3**, 1953–1976 (2017)
9. Bachar, M., Khamsi, M.A., Kozlowski, W.M., Bounkhel, M.: Common fixed points of monotone Lipschitzian semigroups in Banach spaces. J. Nonlinear Sci. Appl. **11.1**, 73–79 (2018)
10. Bachar, M., Khamsi, M.A., Kozlowski, W.M., Bounkhel, M.: Common fixed points of monotone Lipschitzian semigroups in hyperbolic metric spaces. J. Nonlinear Convex Anal. **19.6**, 987–994 (2018)
11. Barbu, V.: Nonlinear Semigroups and Differential Equations in Banach Spaces. Noordhoff, Leyden (1976)
12. Bellomo, N., Brzezniak, Z., Socio, L.M.: Nonlinear Stochastic Evolution Problems in Applied Sciences. Math. Appl., 82. Kluwer Academic Publishers Group, Dordrecht (1992)
13. Belluce, L.P., Kirk, W.A.: Fixed-point theorems for families of contraction mappings. Pacific. J. Math., **18**, 213–217(1966)
14. Belluce, L.P., Kirk, W.A.: Nonexpansive mappings and fixed-points in Banach spaces. Illinois. J. Math., **11**, 474–479 (1967)
15. Blessing, J., Kupper, M.: Nonlinear Semigroups Built on Generating Families and their Lipschitz Sets. Potential Analysis, **59**, 857–895 (2023)
16. Blessing, J., Kupper, M., Nendel, M.: Convex monotone semigroups and their generators with respect to Γ-convergence. J. Funct. Anal., **288**, paper no. 110841 (2025)

17. Browder, F.E.: Nonexpansive nonlinear operators in a Banach space. Proc. Nat. Acad. Sci. USA, **54**, 1041–1044 (1965)
18. Bruck, R.E.: A common fixed point theorem for a commuting family of nonexpansive mappings. Pacific. J. Math., **53**, 59–71 (1974)
19. Bruck, R.E.: A simple proof of the mean ergodic theorem for nonlinear contractions in Banach spaces. Israel. J. Math., **32**, 107–116 (1979)
20. Bruck, R.E., Kuczumow, T., Reich, S.: Convergence of iterates of asymptotically nonexpansive mappings in Banach spaces with the uniform Opial property. Colloq. Math., **65.2**, 169–179 (1993)
21. Butnariu, D., Reich, S., Zaslavski, A.J.: Convergence to Fixed Points of Inexact Orbits of Bregman-Monotone and of Nonexpansive Operators in Banach Spaces. Fixed Point Theory and Its Applications. Yokohama Publishers: Yokohama, Japan, 2006, 11–32.
22. Butnariu, D., Reich, S., Zaslavski, A.J.: Asymptotic behavior of inexact orbits for a class of operators in complete metric spaces. J. Appl. Anal. **13**, 1–11 (2007)
23. Carmona, R., Delarue, F.: Forward–backward stochastic differential equations and controlled McKean–Vlasov dynamics. Ann. Probab., **43**, 2647–2700 (2015)
24. Chen, S.: Geometry of Orlicz Spaces. Dissertat. Math., **356**, 4–205 (1996)
25. Chow, P-L.: Stochastic partial differential equations. Advances in Applied Mathematics, CRC Press, Boca Raton, London, New York (2015)
26. Crandall, M.G., Pazy, A.: Semigroups of nonlinear contractions and dissipative sets. J. Funct. Anal. **3**, 376–418 (1969)
27. Crandall, M.G., Liggett, T.M..: Generation of semi-groups of nonlinear transformations on general Banach spaces. Amer. J. Math. **93.2**, 265–298 (1971)
28. Dawson, D.A.: Stochastic evolution equations. Math. Biosci. **15**, 287–316 (1972)
29. Dehaish, B.A.B., Khamsi, M.A.: Approximating common fixed points of semigroups in metric spaces. Fixed Point Theory Appl. **2015:51** (2015)
30. Dehaish, B.A.B., Khamsi, M.A.: Browder and Göhde fixed point theorem for monotone nonexpansive mappings. Fixed Point Theory Appl., **2016:20**, (2016).
31. Dehaish, B.A.B., Khamsi, M.A., Kozlowski, W.M.: Common fixed points for pointwise Lipschitzian semigroups in modular function spaces. Fixed Point Theory Appl. **2013:214** (2013)
32. Dehaish, B.A.B., Khamsi, M.A., Kozlowski, W.M.: On the convergence of iteration processes for semigroups of nonlinear mappings in modular function spaces. Fixed Point Theory Appl. **2015:3** (2015)
33. DeMarr, R.E.: Common fixed-points for commuting contraction mappings. Pacific. J. Math., **13**, 1139–1141 (1963)
34. Denk, R., Kupper, M., Nendel. M.: A semigroup approach to nonlinear Lévy processes. Stochastic Process Appl. **130.3**, 1616–1642 (2020)
35. Engel, K-J.; Nagel, R.: One-Parameter Semigroups for Linear Evolution Equations. Springer, New York (1999)
36. Engel, K-J.; Nagel, R.: A Short Course on Operator Semigroups. Springer, New York (2006)
37. Espinola, R., Wiśnicki. A.: The Knaster-Tarski Theorem versus monotone nonexpansive mappings. Bull. Pol. Acad. Sci. Math. **66.1**, 1–7 (2018)
38. El-Sayed, S.M., Ran, A.C.M.: On an iteration method for solving a class of nonlinear matrix equations. SIAM J. Matrix Anal. Appl., **23.3**, 632–645 (2002)
39. De Prato, G., Zabczyk, J.: Stochastic equations in infinite dimensions. Cambridge Univ. Press, Cambridge, UK (2014)
40. Goebel, K., Kirk, W.A.: A fixed point theorem for asymptotically nonexpansive mappings. Proc. Am. Math. Soc., **35.1**, 171–174 (1972)
41. Goebel, K., Kirk, W.A.: Topics in Metric Fixed Point Theory, Cambridge Univ. Press, Cambridge (1990)
42. Goebel, K., Reich, S.: Uniform Convexity, Hyperbolic Geometry and Nonexpansive Mappings. Series of Monographs and textbooks in Pure and Applied Mathematics, vol. 83, Dekker, New York (1984).

References

43. Gornicki, J.: Weak convergence theorems for asymptotically nonexpansive mappings in uniformly convex Banach spaces. Comment. Math. Univ. Carolin., **30.2**, 249 – 252 (1989)
44. Hussain, N., Khamsi, M.A.: On asymptotic pointwise contractions in metric spaces. Nonlinear Analysis, **71.10**, 4423–4429 (2009)
45. Ishikawa, S.: Fixed points and iteration of a nonexpansive mapping in a Banach spaces. Proc. Amer. Math. Soc. **59.1**, 65–71 (1976)
46. Jachymski, J.: The Contraction Principle for Mappings on a Metric Space with a Graph. Proc. Amer. Math. Soc. **136.4** 1359–1373 (2008)
47. Kato, T.: Nonlinear semigroups and evolution equations. J. Math. Soc. Japan, **19**, 508–520 (1967)
48. Kim, G.E, Takahashi, W.: Approximating common fixed points of nonexpansive semigroups in Banach spaces. Sci. Math. Jpn., **63.1**, 31–36 (2006)
49. Kirk, W.A.: Mappings of generalized contractive type. J. Math. Anal. Appl., **32**, 567–572 (1970)
50. Kirk, W. A., Xu, H-K.: Asymptotic pointwise contractions. Nonlinear Analysis, **69.12**, 4706–4712 (2008)
51. Khamsi, M.A.: Nonlinear semigroups in modular function spaces. Math. Japon., **37.2**, 291–299 (1992)
52. Khamsi, M.A., Khan, A.R.: On monotone nonexpansive mappings in $L_1([0, 1])$. Fixed Point Theory Appl., **2015:94** (2015).
53. Khamsi, M.A., Kirk, W.A.: An Introduction to Metric Spaces and Fixed Point Theory. John Wiley, New York (2001)
54. Khamsi, M.A.; Kozlowski, W.M.: Fixed Point Theory in Modular Function Spaces. Springer Cham, Heidelberg, New York, Dordrecht, and London (2015)
55. Khamsi, M.A., Kozlowski, W.M., Reich, S.: Fixed point theory in modular function spaces. Nonlinear Anal. **14.11**, 935–953 (1990)
56. Kolokoltsov, V.N.: Nonlinear Markov Processes and Kinetic Equations. Cambridge University Press (2010)
57. Kolokoltsov, V.N.: The Lévy-Khintchine type operators with variable Lipschitz continuous coefficients generate linear or nonlinear Markov processes and semigroups. Probab. Theory Relat. Fields, **151**, 95–123 (2011).
58. Kozlowski, W.M.: Modular Function Spaces.; Monogr. Textbooks Pure Appl. Math., 122, Dekker, New York, NY, USA; Basel, Switzerland (1988)
59. Kozlowski, W.M.: Common fixed points for semigroups of pointwise Lipschitzian mappings in Banach spaces. Bull. Austr. Math. Soc., **84.3**, 353–361 (2011)
60. Kozlowski, W.M.: On the existence of common fixed points for semigroups of nonlinear mappings in modular function spaces. Comment. Math. **51.1**, 81–98 (2011)
61. Kozlowski, W.M.: Fixed point iteration processes for asymptotic pointwise nonexpansive mappings in Banach spaces. J. Math. Anal. Appl., **377.1**, 43–52 (2011)
62. Kozlowski, W.M.: On the construction of common fixed points for semigroups of nonlinear mappings in uniformly convex and uniformly smooth Banach spaces. Comment. Math. **52.2**, 113–136 (2012)
63. Kozlowski, W.M.: On Nonlinear Differential Equations in Generalized Musielak-Orlicz Spaces. Comment. Math., **53.2**, 113–133 (2013)
64. Kozlowski, W.M.: Strong convergence of implicit iteration processes for nonexpansive semigroups in Banach spaces. Comment. Math. **54.2**, 203–208 (2014)
65. Kozlowski, W.M.: On common fixed points of semigroups of mappings nonexpansive with respect to convex function modulars. J. Nonlinear Convex Anal. **15.3**, 437–449 (2014)
66. Kozlowski W.M.: On the Cauchy Problem for the Nonlinear Differential Equations with Values in Modular Function Spaces. In: Shahid, M.H., Ahmad, S, et al. (eds.): Differential Geometry, Functional Analysis and Applications, Narosa Publishing House, New Delhi, 75–105, (2015)
67. Kozlowski, W.M.: On convergence of iteration processes for semigroups in uniformly convex and uniformly smooth Banach spaces. J. Math. Anal. Appl. **426.2**, 1182–1191 (2015)

68. Kozlowski, W.M.: Monotone Lipschitzian semigroups in Banach spaces. J. Aust. Math. Soc. **105.3**, 417–428 (2018)
69. Kozlowski, W.M.: On the construction algorithms for the common fixed points of the monotone nonexpansive semigroups. J. Nonlinear Convex Anal. **20.10**, 2119–2131 (2019)
70. Kozlowski, W.M.: Contractive semigroups in topological vector spaces, on the 100th anniversary of Stefan Banach's Contraction Principle. Bull. Austr. Math. Soc., **108.2**, 331–338 (2023)
71. Kozlowski, W.M.: Modular version of Goebel-Kirk Theorem. Topol. Methods Nonlinear Anal. **63.1**, 99–114 (2024)
72. Kozlowski, W.M.: On stability of iteration processes convergent to stationary points of semigroups of nonlinear operators in metric spaces. Optimization, 1–14, https://doi.org/10.1080/02331934.2024.2410259(2024)
73. Kozlowski, W.M.: Convergence of Implicit Iterative Processes for Semigroups of Nonlinear Operators Acting in Regular Modular Spaces. Mathematics 2024, **12**(24), 4007 (2024)
74. Kozlowski, W.M.: Stability of iteration processes weakly convergent to stationary points of semigroups of nonlinear operators. Rend. Circ. Mat. Palermo, II. Ser, **74.1** Paper No 51 (2025)
75. Kozlowski, W.M.: Common fixed points of semigroups of nonlinear operators in regular modular spaces. J. Nonlinear Convex Anal. **26.5**, 1505–1522 (2025)
76. Kozlowski, W.M.: A note on implicit iteration processes. Bull. Austr. Math. Soc., 1–12, https://doi.org/10.1017/S0004972725000292(2025)
77. Kozlowski, W.M., Sims, B.: On the convergence of iteration processes for semigroups of nonlinear mappings in Banach spaces. In: Bailey, D.H., Bauschke, H.H., Borwein, P., Garvan, F., Thera, M., Vanderwerff, J.D., Wolkowicz, H. (eds.): Computational and Analytical Mathematics. Springer Proceedings in Mathematics and Statistics, vol. 50. Springer, New York (2013)
78. Krasnosel'skii, M.A.: Two remarks on the method of successive approximation. Uspiehi Mat. Nauk, **10**, 123–127 (1955)
79. Kühnemund, F.: A Hille-Yosida Theorem for Bi-continuous Semigroups. Semigroup Forum, **67**, 205–225 (2003)
80. Li, G., Sims, B.: τ-Demiclosedness principle and asymptotic bahavior for semigroup of nonexpansive mappings in metric spaces. Yokohama Publ., Yokohama (2008)
81. Lim, T.C.: A fixed point theorem for families of nonexpansive mappings. Pacific. J. Math., **53**, 487–493 (1974)
82. Mann, W.R.: Mean value methods in iteration. Proc. Amer. Math. Soc., **4**, 506–510 (1953)
83. McKean, H. P., Jr.: A class of Markov processes associated with nonlinear parabolic equations. Proc. Nat. Acad. Sci. USA, **56**, 1907–1911 (1966)
84. Nieto, J.J., Rodriguez-Lopez, R.: Contractive mapping theorems in partially ordered sets and applications to ordinary differential equations. Order **22.3**, 223–239 (2005)
85. Nisio, M.: On a non-linear semi-group attached to stochastic optimal control. Publ. Res. Inst. Math. Sci. **12.2**, 513–537 (1976/77)
86. Neumann, B.A.: Nonlinear Markov chains with finite state space: invariant distributions and long-term behaviour. J. Appl. Probab. **60.1**, 30–44 (2023)
87. Opial, Z.: Weak convergence of the sequence of successive approximations for nonexpansive mappings. Bull. Amer. Math. Soc., **73**, 591–597 (1967)
88. Ostrowski, A.M.: The round-off stability of iterations. Z. Angew. Math. Mech., **47**, 77–81 (1967)
89. Pazy, A.: Semigroups of Linear Operators and Applications to Partial Differential Equations. Springer-Verlag, Berlin, Heidelberg, New York, Tokyo (1983)
90. Pham, H., Wei, X.: Dynamic programming for optimal control of stochastic McKean–Vlasov dynamics. SIAM J. Control Optim., **55.2**, 1069–1101 (2017)
91. Ran, A.C.M., Reurings, M.C.B.: A fixed point theorem in partially ordered sets and some applications to matrix equations. Proc. Amer. Math. Soc. **132.5**, 1435–1443 (2004)
92. Reich, S.: A note on the mean ergodic theorem for nonlinear semigroups. J. Math. Anal. Appl. **91.2**, 547–551 (1983)

93. Reich, S., Shoikhet, D.: Nonlinear semigroups, fixed points, and geometry of domains in Banach spaces. Imperial College Press, London (2005)
94. Reich, S., Zaslavski, A.J.: Convergence of inexact iterates of uniformly locally nonexpansive mappings with summable errors. JP J. Fixed Point Theory Appl., **18**, 1–11 (2022)
95. Reich, S., Zaslavski, A.J.: Three Convergence Results for Inexact Iterates of Uniformly Locally Nonexpansive Mappings. Symmetry, **15.5**, 1084 (2023)
96. Saejung, S.: Strong Convergence Theorems for Nonexpansive Semigroups without Bochner Integrals. Fixed Point Theory Appl., **732008:745010** (2008)
97. Salisu, S., Berinde, V., Sriwongsa, S., Kumam, P.: On approximating fixed points of strictly pseudocontractive mappings in metric spaces. Carpathian J. Math., **40.2**, 419–430 (2024)
98. Schu, J.: Weak and strong convergence to fixed points of asymptotically nonexpansive mappings. Bull. Austral. Math. Soc. **43.1**, 153–159 (1991)
99. Suzuki, T.: On strong convergence to common fixed points of nonexpansive mappings in Hilbert spaces. Proc. Amer. Math. Soc. **131.7**, 2133–2136 (2003)
100. Suzuki, T.: Strong convergence of Krasnoselskii and Mann's type sequences for one-parameter nonexpansive semigroups without Bochner integrals. J. Math. Anal. Appl. **305.1**, 227–239 (2005)
101. Suzuki, T.: Common fixed points of one-parameter nonexpansive semigroups. Bull. London Math. Soc. **38.6**, 1009–1018 (2006)
102. Taiwo, A., Reich, S.: Bounded perturbation resilience of a regularized forward-reflected-backward splitting method for solving variational inclusion problems with applications. Optimization **73.7**, 2089–2122 (2024)
103. Takahashi, W.: Nonlinear Functional Analysis. Yokohama Publishers, Yokohama (2000)
104. Thong, D.V.: An implicit iteration process for nonexpansive semigroups. Nonlinear Anal. **74.17**, 6116–6120 (2011)
105. Tan, K-K., Xu, H-K.: Fixed point iteration processes for asymptotically nonexpansive mappings, Proc. Amer. Math. Soc. **122.3**, 733–739 (1994)
106. Xu, H-K.: Inequalities in Banach spaces with applications. Nonlinear Anal., **16.12**, 1127–1138 (1991)
107. Xu, H-K.: Existence and convergence for fixed points of asymptotically nonexpansive type. Nonlinear Anal., **16.12**, 1139–1146 (1991)
108. Xu, H-K.: A strong convergence theorem for contraction semigroups in Banach spaces. Bull. Austral. Math. Soc. **72.3**, 371–379 (2005)
109. Zaslavski, A.J.: Approximate solutions of common fixed point problems. Springer Optimization and Its Applications 112. Springer Nature, Cham, Switzerland (2016)
110. Zaslavski, A.J.: Algorithms for solving common fixed point problems. Springer Optimization and Its Applications 132. Springer Nature, Cham, Switzerland (2018)
111. Zaslavski, A.J.: Two Convergence Results for Inexact Orbits of Nonexpansive Operators in Metric Spaces with Graphs. Axioms, **12.10**, 999 (2023)
112. Zaslavski, A.J.: Solutions of Fixed Point Problems with Computational Errors. Springer Optimization and Its Applications 210. Springer Nature, Cham, Switzerland (2024)
113. Zaslavski, A.J.: The Krasnoselskii-Mann Method for Common Fixed Point Problems. In: SpringerBriefs in Optimization. Springer, Cham, Switzerland (2025)
114. Zeidler, E.: Nonlinear Functional Analysis and its Applications I: Fixed-Point Theorems, Springer-Verlag, New York, Berlin, Heidelberg, Tokyo (1986)

Index

Symbols
$F(T_t)$, 2
$F(\mathcal{T})$, 2
$K(t)$, 116
$KIS(\mathcal{T}, \lambda, \{t_n\}, x_0)$, 104
$M(t)$, 117
$M_n(t)$, 117
$P(C, \mathcal{T}, x_0, \{c_k\}, \{t_k\})$, 72
$P_{ij}(t, m)$, 130
$P_{k,w}$, 72
S_k^n, 84
$S_\tau(v)(t)$, 116
$\mathbb{N}_0$, 25
$\mathcal{ACS}(C)$, 7
$\mathcal{ANS}(C)$, 7
$\mathcal{APCS}(C)$, 7
$\mathcal{APNS}(C)$, 7
$\mathcal{A}(C)$, 34
$\mathcal{A}_c(C)$, 35
$\mathcal{NE}(C)$, 7
$\mathcal{P}(S)$, 129
$\mathcal{PNE}(C)$, 7
$gI(\mathcal{T}, \{c_k\}, \{d_k\}, \{t_k\})$, 59
$gI_{UB}(\mathcal{T}, \{c_k\}, \{d_k\}, \{t_k\})$, 65
$gKM(\mathcal{T}, \{c_k\}, \{t_k\})$, 42
$gKM_{UB}(\mathcal{T}, \{c_k\}, \{t_k\})$, 49
$mAPNSC(C)$, 104
$\Gamma(\{H_k\})$, 88

A
Additive semigroup $\mathbb{N}_0$, 25
Approximate fixed-point sequence, 27
Asymptotic contractive semigroup, 7
Asymptotic nonexpansive semigroup, 7
Asymptotic pointwise contractive semigroup, 7
Asymptotic pointwise nonexpansive semigroup, 7
Asymptotic pointwise nonexpansive sequence, 33

B
Banach Fixed Point Principle, 15
Bochner integral, 116
Boltzmann equation, 129
Borel measure, 129
Browder Fixed Point Theorem, 8, 20
Bruck lemma, 15

C
Cauchy problem, 115
Chapman-Kolmogorov equation, 129, 130
Common fixed point, 2
Continuous semigroup, 2

D
Demiclosedness Principle, 29
Dynamical process, 3, 115, 129
Dynamical system, 3, 115, 128

E
Eberlein-Šmulian Theorem, 13
Equicontinuous semigroup, 3
Espinola-Wiśnicki Theorem, 113
Eventually nonexpansive semigroup, 7
Evolution equation, 3, 129

F
Finitely generated semigroup, 25
Fréchet differentiable mapping, 5, 20

G
Generating set, 25
Goebel-Kirk example, 9
Goebel-Kirk Theorem, 9

H
Hyperbolic space, 131

I
Implicit iteration process, 71
Initial value problem, 115
Invariant distribution, 130
Ishikawa iteration processes, 59
Iter($\{H_k\}, x$), 85

K
Kirk lemma, 5
Knaster-Tarski Theorem, 101, 113
Krasnosel'skii-Mann iteration process, 41

L
Landau-Fokker-Planck equation, 129
Lévy measure, 129
Locally uniformly bounded Ishikawa process, 65
Locally uniformly bounded Krasnosel'skii-Mann process, 49

M
Markov semigroup, 129
Mazur Theorem, 13, 87, 105
McKean equation, 129
Milman-Pettis Theorem, 13
Modular function space, 131
Modulus of convexity, 8
Monotone closed set, 100
Monotone Krasnosel'skii-Ishikawa iteration process, 104
Monotone mapping, 96
Monotone norm, 98
Monotone Opial property, 98
Monotone pointwise contraction, 97
Monotone pointwise Lipschitzian semigroup, 97
Monotone pointwise nonexpansive mapping, 97
Monotone solution to differential equation, 124
Multi-step iteration process, 58

N
Nonexpansive semigroup, 7
Nonlinear Markov chain, 129
Nonlinear Markov process, 4, 129
Nonlinear parabolic equation, 129
Normalised duality map, 24

O
Opial property, 32
Ordered Banach space, 124
Order interval, 96
Order-preserving mapping, 96

P
Parallelogram inequality, 17
Partial order, 96
Picard iteration process, 15
Pointwise eventually nonexpansive semigroup, 7
Pointwise Lipschitzian mapping, 5
Pointwise Lipschitzian semigroup, 6
Probability simplex, 129

R
Regular sequence, 27
Resolving semigroup, 3

S

Smoluchowski equation, 129
Stability of implicit iteration process, 93
Stability of Ishikawa process, 91
Stability of Krasnosel'skii-Mann processes, 89
Stability under summable errors, 88
Stochastic process, 3, 128
Strong ergodicity, 130
Strongly continuous semigroup, 2
Suzuki's representation of fixed point sets, 58
Symmetry property, 97

T

Type function, 13

U

Uniformly convex Banach space, 8
Uniformly smooth Banach space, 24
Urysohn operator, 131

V

Vlasov equation, 129

W

Well-controlled sequence of pointwise Lipchitzian operators, 85
Well-defined Ishikawa process, 59
Well-defined Krasnosel'skii-Mann process, 42
Well-defined monotone Krasnosel'skii-Ishikawa process, 106

X

Xu's characterisation of uniformly convex Banach spaces, 17

MIX
Papier aus verantwortungsvollen Quellen
Paper from responsible sources
FSC® C105338

If you have any concerns about our products,
you can contact us on
ProductSafety@springernature.com

In case Publisher is established outside the EU,
the EU authorized representative is:
**Springer Nature Customer Service Center GmbH
Europaplatz 3, 69115 Heidelberg, Germany**

Printed by Libri Plureos GmbH
in Hamburg, Germany